KEEPING
BEES

A Handbook for the Hobbyist Beekeeper

BY
FRANKLIN H. CARRIER

ISBN: 0-9607550-1-2

© 1984 By Franklin H. Carrier. All rights reserved. No part of this book may be reproduced, stored in a retrieval system, or transcribed by any means—electronic, photocopying, recording, or otherwise—without the prior written permission of the publisher, Franklin H. Carrier.

Printed in the United States of America

**Published by Carrier's Beekeeping Supplies
601 South Baywood Avenue
San Jose, California 95128**

PREFACE

Keeping Bees—A Handbook For the Hobbyist Beekeeper, covers the second and following years of beekeeping and was written for hobbyist beekeepers who have kept bees for at least one year. It is a companion volume of *Begin to Keep Bees—A Step-by-Step Guide for the Beginning Beekeeper*, which was written for the beginning hobbyist beekeeper and covers the first year of beekeeping.

Keeping Bees can be used year after year by the busy hobbyist beekeeper as a reminder of what activities and manipulations should be performed during the beekeeping year, how they should be performed, and about when they should be performed.

Because *Keeping Bees* presumes that the hobbyist beekeeper has had at least one year of beekeeping experience, it gets right to the subject as quickly as possible. If you are a hobbyist beekeeper who has had bees for several years but have not yet kept bees properly, you might want to obtain a copy of *Begin to Keep Bees*, in which the first year of beekeeping is covered in considerable detail.

The format of *Keeping Bees* is similar to that of *Begin to Keep Bees*, with one important exception. This volume will indicate by the headings "In the South:" or "In the North:" those areas in a unit where different activities or manipulations are required as a result of the considerable variations of temperature occurring in the early spring and again in the fall. Knowing the geographical and climatic area in which you live, you can omit or modify those items discussed or those steps in the directions that are not relevant in your area. You may find it interesting to know the problems faced by beekeepers in other areas. This handbook also will make it easier to reorient yourself to beekeeping if you should move from a northern area to a southern area or vice versa. The uncaptioned photographs found in the book are for the reader's enjoyment of seeing the bees caught in some of their many and varied activities.

The table of contents indicates by the notation ☐ following the section listing those activities or manipulations for which step-by-step directions are given. This enables you to go directly to the steps in a particular segment if directed to do so in any other unit of the text.

Unit 2, "An Overview of the Beekeeping Year," sets the plan for the units that follow by outlining the activities and manipulations performed during the beekeeping year, from late winter, through the beekeeping season, and back to preparing the bees for winter. The general outline of approximate dates is only that, and should not be taken as gospel. It is always best to talk to and ask advice of experienced beekeepers in your area regarding dates, particularly in the South. The manipulative steps in the book will apply when the time is right. Beekeepers using this handbook in the Southern Hemisphere will have to reverse the headings for the North and the South. The steps for the manipulations in the Southern Hemisphere remain the same; the change is the time of year the manipulations are performed.

The final units, under the general heading of "Miscellaneous Activities and Manipulations," are concerned with activities and manipulations that are not necessarily a part of the normal beekeeping year. These may become a necessary part of your activities and manipulations because of special problems with your hive, to perform the manipulations to recover and transfer a colony of bees from strange or odd wild hives, to gather pollen, to set up an observation hive, etc.

I have begun a third volume in this series that will discuss and define tips and tricks in beekeeping. Like grammar and spelling, beekeeping has a number of exceptions which may allow the beekeeper to do things during one part of the year that could or should not be done at other times of the year. Being advised by someone with thirty-five years of beekeeping experience may help you learn more quickly those tips and tricks that will make your beekeeping safer, more pleasant, and less work and will alert you to developing some tricks of your own.

The author wishes to thank Lawrence Pollard, Joseph Fenwick, and Helynn Carrier, whose photographs appear in this book. I also wish to thank my friend, Mary Forkner, owner of Publication Alternatives, who has helped edit the manuscript; obtained electronic typesetting, printing, and binding; and kept track of all the minutiae connected with publishing a book. Many thanks are due to Laura Kenney and Suzy Blackaby who have edited this manuscript, entered the embedded commands for the electronic typesetting to work properly, and then proofread the manuscript again to make sure the commands were entered properly. Again I would like to thank my beloved wife and business partner Helynn for her encouragement, photographic work, and several proofreadings of the manuscript. Helynn has not had to type and retype this manuscript, because I typed it on a word processor and corrections to the manuscript were made using that medium.

This manuscript was typeset by Graphic Typesetting Service in Century Expanded type. The book was printed and bound by Alpine Press. The jacket cover was designed and prepared for color printing by Michael Rogondino from a photograph by Helynn Carrier. Photographs in this book were made from 5 × 7 enlargements prepared by Process Techniques of Sunnyvale, Ca. and 60 Minute Memories of Saratoga, Ca. This book was published by Carrier's Beekeeping Supplies, which is owned and operated by Franklin and Helynn Carrier.

Franklin H. Carrier

CONTENTS

PART I	AN OVERVIEW OF THE BEEKEEPING YEAR	1
Unit 1	INTRODUCTION 3	

A Time to Wait 3
Some Things You Can Do While Waiting for Spring 4

Unit 2	AN OVERVIEW OF THE BEEKEEPING YEAR 6	
PART II	MIDWINTER PLANNING AND ACTIVITIES	9
Unit 3	ORDERING A PACKAGE OF BEES	11

To Order a Package of Bees ☐ 11
Ordering a Package of Bees to Restock a Dead Hive ☐ 13

Unit 4 ORDERING QUEENS 15

Ordering Queens for Spring Re-Queening ☐ 15
Ordering Queens for Midseason Re-Queening 17
Ordering Queens for Fall Re-Queening 17

Unit 5 ORDERING HIVE EQUIPMENT 18

A Producing Beehive 18

Selecting Honey Supers 20
Types of Honey You Can Produce 20
Where to Buy Hive Equipment 21
Words of Caution When Ordering Wax
 Foundation 22
Ordering Hive Materials for Existing Hives ☐ 22
  Deep Supers for Honey Storage 22
  Medium Supers for Honey Storage 23
  Shallow Supers for Honey Storage 23
  Section Comb Supers for Honey Storage 24
  Supers of Mixed Sizes for Honey Storage 24
Ordering Hive Materials to Start New Hives ☐ 24
Ordering Miscellaneous Supplies for Existing
 Hives ☐ 26
You Can Save Money if You Buy in Quantity ☐ 27

Unit 6   **MIDWINTER HIVE ACTIVITIES AND MANIPULATIONS 29**

PART III **EARLY SPRING ACTIVITIES AND MANIPULATIONS   31**

Unit 7   **A PREVIEW OF EARLY SPRING ACTIVITIES AND MANIPULATIONS 33**

Some Reasons Your Bees Might Not Have Made It
 Through the Winter 35

Unit 8   **ABOUT BEE DISEASES 38**

Learn to Recognize the Foulbrood Diseases (AFB/
 EFB) 38
American Foulbrood (AFB) 38
  Controlling American Foulbrood 39
European Foulbrood (EFB) 40
  Controlling European Foulbrood 40
Comparison Characteristics of the Foulbrood
 Diseases 40
Nosema Disease 41
  Controlling Nosema Disease 42
Diagnosis of Bee Diseases 42

Unit 9 A REVIEW OF SOME IMPORTANT ACTIVITIES AND MANIPULATIONS 43

Explanation of Step-by-Step Procedures 43
Materials for Dressing Properly to Work the Bees 43
Dress Properly ☐ 45
Smoking the Hive 46
   A Smoker Alternative 47
   The Smoker 48
   Smoker Fuels 48
Light the Smoker ☐ 48
Enter the Hive ☐ 49
Locating the Hive 50
Moving a Hive of Bees ☐ 51
Removing a Bee Stinger 53

Unit 10 THE PREMEDICATION CHECK 55

Performing the Premedication Check ☐ 55
Some Tips on Checking a Dead or Dying Hive for Foulbrood 58

Unit 11 EARLIEST MANIPULATIONS OF THE HIVE 60

About Terramycin®, Which Controls AFB/EFB 60
   Preparing the AFB/EFB Medication 61
Pollen Supplement 62
   Preparing the Pollen Supplement Patty ☐ 63
Medicating to Control AFB/EFB and Feeding the Pollen Patty ☐ 63
Preparing the Sugar Solutions for Feeding and Medicating ☐ 65
Medicating to Control Nosema Disease ☐ 66
Setting Up a Constantly Available Water Source 66

PART IV SPRING ACTIVITIES AND MANIPULATIONS 69

CONTENTS   ix

Unit 12   INSPECTING THE BROOD
            CHAMBER   71
            Inspect the Brood Chamber ☐   77

Unit 13   RE-QUEENING THE HIVE   81
            De-Queening the Hive   83
                Searching the Frames for the Queen ☐   83
                Dumping the Bees to Find the Queen   85
                    Shaking the Bees from the Frames ☐   85
                    Why Dumping Works   85
                Dumping the Bees to De-Queen the
                    Hive ☐   86
            Re-Queen the Hive ☐   88
                After Four Days ☐   91
                After Seven Days ☐   94

Unit 14   INSTALLING A PACKAGE OF
            BEES   95
            When the Package of Bees Arrives   95
            Feeding the Package of Bees When You Get Them
                Home   96
                    Preparing the Sugar Solution to Feed the
                        Bees ☐   96
                    Feeding the Package of Bees ☐   96
            Holding a Package of Bees for Several Days, if
                Necessary   96
            Gorging the Bees Before Installing Them in the
                Hive ☐   97
            Removing the Queen Cage from the Package of
                Bees ☐   98
                    Queen Cages in Package Bees   99
            Methods for Installing the Package of Bees   99
                Method 1 ☐   100
                Method 2 ☐   104
                Method 3 ☐   107
                About a Dead Queen   111
                After Four Days   112
                    Method 1 ☐   112
                    Method 2 ☐   113

x CONTENTS

    Method 3 ☐ 115
     After Seven Days 116
     First Inspection of the Brood Chamber ☐ 116

Unit 15 SWARMS AND SWARM CONTROL TECHNIQUES 118

  Why Bees Swarm 118
  A Discussion of Swarming 118
  Why the World Is Not Overpopulated by Honey Bees 123
  Multiple Swarms from the Same Hive 124
  Queen Substance 125
  Some Facts about Swarms and Swarming 126
  Some Swarm Control Methods that Might Work 127
  Step-by-Step Directions for Some Swarm Control Methods 131
   Re-Queening Each Fall or Spring 131
   Making the Man-made Swarm or Making a Divide ☐ 131
    The Man-made Swarm After Twenty-four Hours ☐ 133
    The Hive Divide After Twenty-four Hours ☐ 133
    Weekly Inspection of the Man-made Swarm Divides ☐ 134
    Uniting the Man-made Swarm Divides ☐ 135
   Moving Frames of Brood above the Queen Excluder ☐ 136
   Switching Locations of Strong Hives and Weak Hives ☐ 137
   Reversing Brood Chambers ☐ 139
   Tearing Out Queen Cells 139
    Finding the Queen Cells 140
    Steps for Tearing Out Queen Cells ☐ 140
   Buying a Queen with a Clipped Wing 142
  If Your Hive Swarms 143

Unit 16 CAPTURING AND HIVING SWARMS 144

  Getting on a List for Swarms 144

Before You Go Out to Capture a Swarm   146
Capturing a Swarm   147
    Capturing a Swarm on a Supple
      Branch ☐   147
    Capturing a Swarm on the Ground ☐   148
    Capturing a Swarm on a Solid Object ☐   149
    Capturing a Swarm on a Large Branch ☐   150
    Capturing a Swarm on a High, Supple
      Branch ☐   151
Hiving a Swarm   152
    Holding a Swarm for More Than Forty-eight
      Hours ☐   152
    Top Dumping to Hive a Swarm ☐   153
    Front Dumping to Hive a Swarm ☐   155
    Inspection After Seven Days   157
    Capturing the Queen When You Front Dump a
      Swarm ☐   158

Unit 17   WHAT TO DO WITH SWARMS   161

Some Reasons Swarms Are Not Utilized by Many
    Beekeepers   162
Methods of Uniting Swarms   163
    Removing the Queen from the Swarm ☐   164
    Carrier's Swarm-Uniting Method ☐   165
    Why This Method Works   167
    Uniting Swarms Using the Newspaper
      Method ☐   167
    What About Swarm Queens?   169
How To Utilize Swarms—Yours and Others   169
Banking Queens   172

PART V   LATE SPRING, SUMMER, AND EARLY FALL ACTIVITIES   173

Unit 18   DURING THE HONEY FLOW   175

The Hive Before the Honey Flow   176
The Hive During the Honey Flow   177
The Hive After the Honey Flow   178

Preparing to Add Honey Supers 179
    Rule of Thumb for Adding Supers (Brood or Honey) 179
    Inspecting the Honey Super 179
    Manipulating the Frames in the Honey Super 180
    Consider These Things When Adding Honey Supers 180
Adding Honey Supers 184
    Adding the First Honey Super and the Queen Excluder ☐ 185
    Inspecting the Honey Super or Pair of Honey Supers ☐ 186
    Adding the Second Honey Super or Pair of Honey Supers ☐ 187
    Adding the Third Honey Super or Pair of Honey Supers ☐ 189
    What to Expect at This Time in the Honey Flow 190
    Why Remove Honey Supers and Extract in Mid-flow 191
    Adding the Fourth and Following Supers ☐ 192
    About the Notes Made After Each Brood and Honey Super Inspection 192

Unit 19   REMOVING HONEY SUPERS 194

Brushing Bees from Frames of Capped Combs ☐ 194
Clearing Bees from Honey Supers with the Bee Escape ☐ 196
    Some Problems You May Encounter Using the Bee Escape 199
Using a Fume Pad and a Chemical Bee-Removing Liquid ☐ 200
Using a High-velocity Blower ☐ 203

Unit 20   PROCESSING AND STORING COMB HONEY 205

Processing Section Comb Honey in Basswood Boxes ☐ 206
Processing Round Plastic Rings of Comb Honey ☐ 207

Processing Comb Honey Cut for Plastic
   Boxes ☐ 207
Processing Comb Honey in the Frame ☐ 208
Processing Chunk Comb Honey ☐ 208
Cleaning Up Sticky Frames After Cutting Out the
   Comb 209
Granulated Comb Honey 209
   Liquefying Granulated Comb Honey ☐ 210
   Liquefying Granulated Chunk Honey 211
   Liquefying Granulated Extracted Honey 212

Unit 21 EXTRACTING HONEY   213

Time to Extract 213
Preparing to Extract 214
Extracting the Honey 218
   Method 1: Extracting for the Small
      Operation ☐ 218
   Method 2: Extracting for the Modest
      Operation ☐ 224
Cleaning the Extracting Equipment ☐ 228

Unit 22 CLEANING AND STORING SUPERS OF EXTRACTED COMB   230

Cleaning Up Supers of Sticky Extracted
   Comb ☐ 231
Storing Cleaned-Up Supers of Extracted
   Comb ☐ 232
   Method 1: Bagging Clean Honey Supers ☐ 233
   Method 2: Stacking Clean Honey
      Supers ☐ 233
Handling Sticky Supers of Extracted Comb 235
Storing a Hive That Died Out 236

PART VI PREPARING FOR WINTER   237

Unit 23 FALL MEDICATION AND FEEDING OF THE HIVE   239

Fall Medicating and Feeding 239
Fall Medicating to Control AFB/EFB ☐ 241

xiv CONTENTS

    Fall Feeding to Control Nosema Disease ☐ 242
      In the North ☐ 242
      In the South ☐ 243
    Fall Feeding of a Non-medicated Sugar Solution 243

Unit 24 UNITING WEAK HIVES 244

    Uniting a Queenless Hive with a Strong Queenright Hive ☐ 246
    Uniting a Weak Queenright Hive with a Strong Queenright Hive ☐ 248

Unit 25 WINTERING THE HIVES 251

    Wintering Hives in the North 252
      Odds and Ends 254
        Emergency Feeding in Late Winter or Very Early Spring 254
    Wintering Hives in the South 255

PART VII MISCELLANEOUS ACTIVITIES AND MANIPULATIONS 257

Unit 26 LAYING WORKERS 259

    What to Do When You Have Laying Workers in Your Hive 262
      If You Have Only One Hive 262
      If You Have Two or More Hives ☐ 262

Unit 27 WAX MOTH INFESTATION 265

    First Signs of Wax Moth Infestation ☐ 266
    Controlling Wax Moth Once Infestation Has Set In 266

Unit 28 REMOVING BEES FROM HOUSES, TREES, ETC. 268

    Access to the Hive 269
    Once Funneling Has Begun 270

Unit 29 REMOVING BEES WHEN THE COMB CAN BE EXPOSED 272

CONTENTS xv

Transferring Bees and Brood From Odd
   Hives ☐ 274
Some Tips on Strange Odd Hives 281

**Unit 30 REARING A FEW QUEENS FOR YOUR
          OWN USE 282**

Raising a Few Queens 283
Queens from Swarm Queen Cells 284
Generating Emergency (Replacement) Queen
   Cells 284

**Unit 31 RE-QUEENING A HIVE OF MEAN
          BEES 287**

De-Queening and Re-Queening a Mean Hive ☐ 288

Unit 32 ROBBING 290

Is It Orientation, Swarming, or Robbing? 290
How to Identify Orientation Flights 290
   Mistaking Orientation Flights for
      Swarming 291
   Mistaking Orientation Flights for Robbing 291
Robbing by Bees 293
   Precautions to Prevent Frantic Robbing 294
   What To Do if Robbing Starts 295

Unit 33 THE OBSERVATION HIVE 297

Some Don'ts Regarding Observation Hives 297
Stocking the Observation Hive 299
   Starting with a Frame of Brood ☐ 299
   Starting with a Small Swarm ☐ 300
Removing the Observation Hive from the
   House ☐ 301
Keeping the Observation Hive from
   Swarming ☐ 303
What You Can See in the Observation Hive 304

Unit 34 POLLEN TRAPS 309

How a Pollen Trap Works 309
Two Types of Pollen Traps 310
Pollen for Human Consumption 313

Pollen for Feeding the Bees 313

Unit 35   **THE SOLAR WAX MELTER**   314

Things To Consider When Buying or Building a Solar Melter 314
Melting Cappings and Scrap Wax 315
Refining Wax for Candle-Making 316
Exchanging Wax Blocks for Foundation 317

Unit 36   **LOOKING BACK AND LOOKING AHEAD**   318

Looking Way Back 318
Looking Some Distance Back 318
Looking Back in the Americas and Australia 319
Looking Ahead 319
The Here and Now 320

GLOSSARY   323

INDEX   328

PART I

AN OVERVIEW OF THE BEEKEEPING YEAR

The next two units are an introduction to the upcoming beekeeping year and a chronological listing of the main activities and manipulations you will perform during the beekeeping year. Unit 2 is a particularly good reference for the busy hobbyist beekeeper.

- **Unit 1** INTRODUCTION
- **Unit 2** AN OVERVIEW OF THE BEEKEEPING YEAR

UNIT 1

Introduction

A TIME TO WAIT

Beekeepers from the equator to as far north and south as bees are kept wait for spring and activity around the hive. From the equator to about 33° N and S latitude, the bees fly almost every day, kept in the hive only by rain and high winds. From 33° to about 60° N and S latitude, winters become increasingly longer and more severe as one proceeds to the higher latitudes. Above 60° N latitude, it becomes almost impossible to winter bees in the open on their own. This does not mean that you cannot keep bees above 60° N latitude; you can, and you may get a good store of honey. At that latitude you will almost certainly have to restock your hive with a new package of bees each spring, because the winters are just so severe that the bees cannot survive. There is very little landmass between 60° and 90° S latitude except the Antarctic.

In the U.S. along the lower half of the eastern coastal range, across the lower tier of states, and along the western coastal plain, winters are relatively mild, so bees will fly almost daily in winter. Across the lower third of the U.S., inland from the coastal plains, and above the lower tier of states, winters are moderate with one or two months when the bees are confined to the hive due to rains and colder weather.

Across the upper two-thirds of the U.S., inland from the west coast but along the upper half of the eastern coastal range, winters are relatively severe with snow and blowing winds. Spring comes even later as you proceed farther north in this area. In the northern tier of states, spring comes slowly in by late March or early April, with an average last frost in spring about May 10. Keep in mind that in the mild or moderate winter areas, in higher elevations, winters can be relatively severe.

Wherever you are you may wonder, "When will spring come?" Later you may wonder, "When will winter come to my area?" To get a fairly good feel for when to expect the beekeeping spring and the beekeeping winter, look in the *Almanac Yearbook*, sold by many newspapers and magazines every January and available in most libraries. Under "Temperature" you will find a listing of average daily temperatures for each month (over the last twenty years) in selected cities throughout the U.S. Also listed for those same cities is the last frost in spring and the first frost in fall, both average dates over the last twenty years.

If the average daily temperature for March in a selected city near you is 55° F, you can assume that a number of days are going to be warmer than that, and you should see some activity at the hive entrance on those days.

SOME THINGS YOU CAN DO WHILE WAITING FOR SPRING

1. Remember, your best source of information about beekeeping in your area will be from experienced beekeepers. Make a beekeeper your friend or take a beekeeper to lunch. Most beginning beekeepers believe 100% of what they hear from other beekeepers and only 10 percent of what they read about beekeeping. *Be selective,* and listen to *experienced* beekeepers who have wintered hives successfully and produced good crops of honey over the years and who keep abreast of the changes and advancements in apiculture.
2. Attend meetings of a beekeeping association, if there is one in your area. Most members are hobbyist, beginning beekeepers and are looking for answers to their questions in the discussions and from the speakers at those meetings. If there is no association in your area but there are interested beekeepers, help get an association started. Meetings usually are on a specific day of each month, last about one and one-half hours, and the meeting place usually can be obtained free or very reasonably from banks, insurance companies, etc.
3. Read beekeeping books. Visit the library and read any new books on beekeeping. You will probably get more out of rereading some of the good books on beekeeping now that you have had some experience. If your library does not subscribe to one of the beekeeping magazines, ask, and maybe they will. Otherwise, subscribe to one or two of them yourself.
4. Write the U.S. Department of Agriculture, Washington, D.C., and ask for Farmers' Bulletin #2255, "Identification and Control of Honey Bee Diseases." Learn to identify American and European Foulbrook and Nosema Disease. Reread this bulletin on the night before

inspecting your brood chambers, and always look for the symptoms of those diseases, particularly American Foulbrood Disease (AFB).
5. Check the hive occasionally. Lift the hive from the rear every two or three weeks to check on the weight loss as the winter proceeds. Keep the auxiliary opening free of snow, frost, and ice. The lift check of hive weight becomes more important as spring approaches. As brood build-up begins, the hive uses honey stores more rapidly. If the hive seems to lighten considerably, plan to feed as soon as the temperature for a part of each day reaches 60° F and the bees become active. Remember that brood rearing begins before the spring flowers start to produce nectar and pollen in almost all areas where bees are kept.
6. Review your notes from last year and plan for the coming year. Plan to continue keeping notes on all your hive inspections, of both brood and honey supers, for the next few years. Soon you will have a good idea of how and when the honey flow occurs, the availability of early and late pollen and nectar sources, when to add supers, and (in northern areas) when to get supers off and extracted to get clean-up and medicating done before cold weather sets in.
7. Make plans for the coming year. Order materials, packages of bees, and queens as soon as your plans are firm. Some aspects of your planning might include:
   - When to medicate?
   - Should you feed?
   - Should you give pollen supplement?
   - What will you do if you lose your hive over the winter?
   - Do you plan to add more hives next year?
   - When and where should you order your package bees?
   - Do you plan to re-queen in the spring? in the fall?
   - Where can you get mated queens and when?
   - What supplies do you need for your existing hives? to start new hives?
   - What are your plans for swarm control?
   - Are you going to plant flowers or shrubs for your bees?
   - Do you want to try a two-queen hive system?
   - Do you want to set up an observation hive?
   - Do you want to trap pollen for spring feeding? for a supplement to your diet?
   - Do you plan to move the hive? When and how?

Most of the preceding items and others are discussed in detail in later units of this volume.

UNIT 2

An Overview of the Beekeeping Year

Before proceeding with any manipulative steps to be performed during the beekeeping year, it might be helpful to review the important hive manipulations and associated activities in the order in which they will occur. The manipulative steps described in this volume will apply in any area of the U.S. (or of the world) where the movable frame hive is used. The timing of the manipulation, particularly the earliest and latest, will depend on the climate in your area.

As this volume proceeds with the early and late manipulations, two areas of the country will be specified: the northern areas and the southern areas.

1. The southern areas: That general area from the lower half of the eastern coastal plain, across the lower tier of states, and up the western coastal plain. In these areas the bees fly almost every day of the year.
2. The northern areas: All areas not included above (in the U.S.). These areas generally have a definite winter season (maybe only one month), including areas with occasional snow and cold weather below 50° F, and areas of rather severe winter weather.

In the northern areas it is easier to specify when to begin the early activities, because the bees will tell you by becoming active when the temperature for a part of each day reaches 60° F or higher. In the southern areas, where bees fly almost every day, you pick a general date about eight to ten weeks before the honey flow begins to start the early activities.

The following table shows the approximate times that the beekeeping activities might be performed in the northern and southern areas. Remember that these are general and should be verified or modified by experienced beekeepers in your area or by experience and notekeeping over several years.

Beekeepers living in the Southern Hemisphere will have to modify this table to reflect the winter in that area. This will mean that the heading *Northern Area* will become *Southern Area*, because the farther south you go in the Southern Hemisphere the more severe the winters become. Under the new heading *Northern Area*, the months will have to be changed to reflect the winter months in the Southern Hemisphere. (For example, October is changed to April, January is changed to July, etc.) Other than the heading changes and the month changes, all the information in the table remains pertinent.

TABLE 2-1 APPROXIMATE DATES FOR HIVE MANIPULATIONS

NORTHERN AREA Daily temperature	*SOUTHERN AREA Day/ Month*	*Manipulation/Activity*
Below 50° F.	Oct. to Jan.	Keep auxiliary exit/vent clear of snow and ice. Heft the hive occasionally to check honey stores available. Order hive equipment, package bees, queens, medicines, pollen supplement, moth crystals. Read and reread beekeeping books. Order seed catalogs. Review notes of last year's hive inspections. Attend beekeepers' association meetings.
Part of each day is 60° F or more.	Jan. 15 to Feb. 15.	Check for hive activity. If none, you may have "winter kill." Medicate to control AFB/EFB without fail. Medicate to control Nosema Disease. Feed pollen supplement. Feed the hive if required—or as required. Set up a water source. Keep it filled all season. Order a package of bees if your hive has died out.
Part of each day is between 65-70° F.	Feb. 1 to March 1.	Check brood supers to see if queen is laying well and in a good pattern. If not, order a queen with delivery as soon as possible. When the packages of bees arrive: (1) start new hives, or (2) restock any hives that have died over the winter. When the queen arrives, re-queen your hive. Feed the hive, if required. Remove hive entrance reducer.

8 KEEPING BEES

Part of each day is 70 to 100° F through to fall.	Feb. 21 through to fall.	Plan your strategy for swarm control. Add the queen excluder and the first honey super (no sooner than ten days after the last AFB medication). Have the hives ready if you plan to start new colonies by capturing swarms. Inspect the honey supers at least every two weeks, preferably more often. Make and keep notes on all brood and honey super inspections. Add the second honey super when required. At this time inspect the brood chamber to make sure your queen is laying and in a good pattern. Check also for signs of AFB. Add the third honey super when required. Add the fourth honey super when required. If you plan to re-queen, order queen for proper delivery date. Add additional honey supers, if and when needed. Remove honey supers, extract, return supers to the hive for clean-up, then remove and store in plastic bags with Paradichlorobenzene (PDB) moth crystals. When your queen arrives, re-queen the hive.
Part of each day is above 60 ° F.	Sept. 15 to Oct. 15	Unite any weak hives that you feel will not make it through the winter. Check the brood chambers to see that the queen is there and laying. Check for AFB/EFB. Insert the hive entrance reducer. Medicate the hive to control AFB/EFB. Medicate to control Nosema Disease. Feed the hive, if necessary, for a better wintering condition.
55° F or below.	Oct. to Jan.	Wrap the hive, if necessary in your area. Be sure to provide an auxiliary exit/vent for the winter. Heft the hive occasionally during the fall and winter for stores check. Review your notes from this and previous years. Make plans for next year. Attend beekeepers' association meetings. Order equipment, package bees, queens, medicines, etc., for next year.

PART **II**

MIDWINTER PLANNING AND ACTIVITIES

The next four units cover those activities not directly associated with the hive in midwinter but best done at this time to assure that you will receive package bees, replacement queens, hive equipment, and miscellaneous supplies before the bees become active in the spring.

- **Unit 3** ORDERING A PACKAGE OF BEES
- **Unit 4** ORDERING QUEENS
- **Unit 5** ORDERING HIVE EQUIPMENT
- **Unit 6** MIDWINTER HIVE ACTIVITIES AND MANIPULATIONS

UNIT 3

Ordering a Package of Bees

The same directions apply if ordering a package of bees to start a new hive this year or to replace a hive that has died out over the winter. The only difference is the time of ordering the package of bees. One does not expect to lose the bees over winter, so it will be spring before you know the hive has died out. (See Figure 3-1.)

TO ORDER A PACKAGE OF BEES

Know these things when ordering package bees:

1. Last spring frost or early fruit bloom in your area.
2. Type of bee you want—Italian, Caucasian, or Carniolan.
3. Where to order package bees. The bee magazines are about the only place that have advertisements for package bees.

☐ Order package bees early. December or January is not too early. Package bee producers sell their packages on a first-come, first-served basis. They begin taking orders in the fall; if their estimated supply of available packages is sold, you must find another producer.

Remember: Beekeepers living east of the Rocky Mountains should select package bee producers east of the Rocky Mountains or pay for air mail delivery. The package of bees will come parcel post, and a trip over the Rocky Mountains can often leave a package of bees in poor shape. Beekeepers living west of the Rocky Mountains should, likewise, order package bees from producers west of the Rocky Mountains. Most package bee producers will not accept orders to cross the Rocky Mountains unless air mail delivery is specified.

Beekeepers living in other areas of the world should consider distance in relation to the time the package of bees will be in transit. Package bees can be safely shipped halfway around the world by air. Try to keep the travel time of package bees under five days.

12 KEEPING BEES

Figure 3-1A, A Package of Bees. A cardboard package of bees with the feeder can removed. The queen has been removed and is being held by the metal strip. (See Figure 14-1.)

Figure 3-1B, Queen Cages from Package Bees. Three types of queen cages that hang in package bees. Left to right: The short queen cage showing the metal strip; the standard wooden queen cage that hangs by a wire attached to the cage; the plastic queen cage that hangs by a plastic strip attached to the cage. Notice the plastic caps on the plastic cage, which open to release the queen.

☐ When ordering package bees, specify whether you want a two-pound or three-pound package, the date you want delivery (about the last frost in your area), and the type of bee you want. Send the total amount for the package or packages. You should receive a card confirming your order, the type of bees to be sent, the approximate shipping date, and a receipt for your payment in full.

If you have no particular preference for a type of bee, you cannot go wrong with Italian or Starline. Do not be surprised if sometime you receive a two- or three-pound package of bees different than the type you ordered. Your queen will be of the correct type. What sometimes happens is that the producer runs out of bees and must buy (healthy) bees from other beekeepers to meet his orders. Any type of bee will get the hive off to a good start, and within about sixty days all your bees will be the offspring of the queen type you ordered.

In the North: A delivery date of about May 1 to May 10 is good.

In the South: Try for the earliest delivery date possible, but a delivery date of about April 1 is usually the best you can expect. Package bees cannot be shipped until the population in the producer's hives has built up in the spring. Bad weather—cold or rain—can affect the delivery date some years.

The three-pound package is best for starting a new hive. In every case, a three-pound package will almost always move ahead faster. If later in the spring you find your hive died out and you need a new package to restart that hive, a two-pound package will do because you will have drawn comb, pollen, and possibly honey stores.

☐ Plan to have your hive on hand, built up, painted, and located before your package of bees arrives.

☐ A week or two before your package of bees is expected to arrive, let your mail carrier know you are expecting a package of bees. Let the carrier know where you can be reached to pick up the package, if necessary. When your bees arrive, they will need to be fed and hived as soon as possible. The procedure and steps for feeding and installing a package of bees in the hive are covered in Unit 14.

ORDERING A PACKAGE OF BEES TO RESTOCK A DEAD HIVE

☐ Follow the steps given previously. Change step one to the following steps, because you will have to order at the time you discover the winter kill, which may be after the temperature reaches 60° F for a part of each day in the spring.

14 KEEPING BEES

☐ Step one changes: (1) Select several package producers handling the type of bees you want. Some may not be able to fill new orders at this time.
(2) Write—better yet, *call*—to see if you can place an order for a package of bees, and find out how soon you can have delivery.

Check to see if the producer honors a major credit card; if you wish, you may be able to order a queen or package and pay by giving the producer your card number.

UNIT 4

Ordering Queens

ORDERING QUEENS FOR SPRING RE-QUEENING

Know these things when ordering queens:
1. Last spring frost or early fruit bloom in your area.
2. Type of queen you want, Italian, Starline, Caucasian, Midnite, or Carniolan.
3. About the only place you can find advertisements for queens is in the bee magazines, which may be available at your library or you can subscribe to one or several.

- ☐ Order the type of queen you want as early as possible. December or January is not too early to make sure you get the queen at the time you choose to re-queen.
- ☐ Your choice of queen type will determine your producer. Some producers handle two types of queens, others handle only one. Starline queens are an Italian hybrid, and Midnite queens are a Caucasian hybrid.
- ☐ Order your queen from any producer in the U.S. handling the type you want. The queens come in light, small cages that are sent by first class mail or air mail, so they arrive within a few days after being sent. Often you must pay for insurance.
- ☐ When ordering, specify the type and number of mated queens you want and the date you want delivery. Send payment in full with the order. You should receive a card confirming the number, type, and approximate delivery date of your queens. It is not necessary to let your mail carrier know that you will be receiving queens in the mail. Queens are mated unless you specify that you want unmated queens.
- ☐ Queens are usually available in the spring after about April 1. How early queens are available will depend on the weather conditions in

15

Figure 4-1, Queen Cages Sent Through the Mail. Three types of queen cages sent through the mail. Left: Two views of the cardboard queen cage. Right: The wooden queen cages. Usually the longer cage is sent, and either would have a candy plug in one of the segments. In all cages sent through the mail, several workers accompany the queen to feed and tend her while in transit.

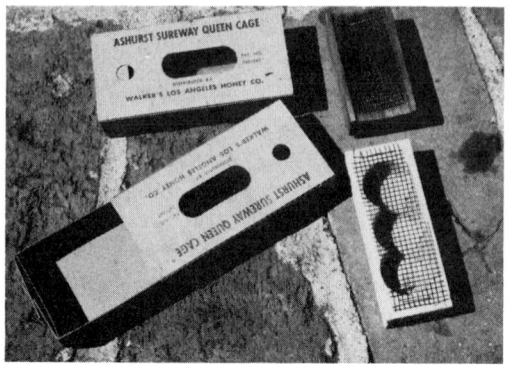

the southern areas where package bees and queens are produced. Unlike package bees, queens become easier to get as the season progresses. Queens may become somewhat less expensive after about June 1 and are readily available until about early September. Some producers "bank" queens into early winter for commercial and hobbyist beekeepers in the southern areas.

☐ Queens are shipped in two types of containers, known as queen cages, with about six worker bees to feed and tend the queen. (See Figure 4-1.)

A. One type of cage is a wooden block about one inch wide, three inches long, and ¾ inch thick. This block is hollowed out, and a small hole is bored on each end of the block leading to the hollow center. A screen covers the cavity in the block; it is in this cavity that the queen and her retinue are placed. One end of the cavity contains a firm sugar candy to feed the bees and queen on the trip and later to act as an automatic release when the cage is placed in the hive for introduction. Small corks block the holes in each end of the cage to keep the queen and her escorts inside until you decide to place the cage in the hive. (See Figures 4-1 and 13-1.)

B. Another type of cage is a thin cardboard container similar to a small matchbox. Inside the box is a small lump of firm sugar candy to feed the bees on the trip. The queen and her escorts are placed inside the box, and the cover is slid over the box and fastened closed. The thin cardboard box has small holes for ventilation. When this type of queen cage is used, it is placed against a frame containing the brood, the cover is slipped off the end of the box, and the box is pressed firmly into the comb. In several days the bees in the hive slowly tear away the cardboard to release the queen. (See Figures 4-1 and 13-3.)

- [ ] When the queen or queens arrive, they can be held for several days if necessary. Keep them out of drafts or direct sunlight. On arrival and once each day, while they are being held in the house, place a drop of water on the screen of the cage or through one of the small holes in the cardboard box.
- [ ] The new queen should be introduced into the hive as soon as possible but *not before* the old queen in the hive has been removed (de-queening).
- [ ] Steps for removing the old queen from the hive (de-queening) and introducing the new queen in the hive (re-queening) are presented and discussed in detail in Unit 13.

ORDERING QUEENS FOR MIDSEASON RE-QUEENING

Usually re-queening is necessary because a virgin queen has been lost on her mating flight (after the hive has swarmed), leaving behind no brood young enough for the bees to replace the queen naturally. If your hive appears to slow down considerably or you see no egg or larvae and no covered brood when you inspect the brood chambers, chances are you have no queen and should order a queen for delivery as soon as possible. Follow the previous steps for ordering a new mated queen.

ORDERING QUEENS FOR FALL RE-QUEENING

In the North: Order your queen so she will arrive about three weeks before—or just as—the honey flow ends. Too early is better than too late. Fall re-queening is probably best if you can have six to eight weeks after re-queening for the new queen to lay a good pattern of brood to produce a good population of young bees for wintering. If you cannot have those six to eight weeks, you would probably be better off to re-queen in the spring.

In the South: Some of the larger package and queen producers in the far South hold or "bank" mated queens, which are available to beekeepers in the southern areas where the bees fly almost every day. These queens may be available almost all winter long. Contact the queen producers in the bee magazines to find who has queens available. In introducing a queen in midwinter, where the night temperatures may fall to as low as 40° F, hang the queen cage so that she will be in the cluster, rather than setting the queen cage on top of the brood frames where she may chill and die if she is not kept warm enough.

UNIT 5

Ordering Hive Equipment

A PRODUCING BEEHIVE (Year Two and Following Years)

To order hive equipment for an existing first-year hive, you should have an idea of what a producing beehive looks like at the end of the second year, before you extract in the fall. (See Figure 5-1.) Knowing this, you can take inventory and order those pieces of equipment you will need. The hive shown in Figure 5-1 consists of the following items:

(1) bottom board
(2) brood chambers (deep supers)
(1) queen excluder
(4) honey supers (deep supers are shown here)
(1) inner cover
(1) hive top

Instead of four deep supers as shown, you *could* use six medium supers *or* eight shallow supers *or* four to six section box supers. You can also mix supers (with the exception of the section box supers) on the hive to come up with the approximate capacity of four deep supers. As a guide, three medium supers equal two deep supers, and two shallow supers equal one deep super.

This may be more than enough storage space. A hive that winters well and gets a good start in the spring will produce more honey the following years than the first year. The author, in California, has averaged three and one-half to four deep supers of honey most years. But in his first year of beekeeping in South Dakota, he started a hive with a three-pound package of Italians on about May 10. By September of that year, the bees had filled four section comb supers with 112 basswood boxes of comb honey. The author also killed the hive that winter by packing it in leaves and straw to keep it warm. He later learned this

ORDERING HIVE EQUIPMENT   19

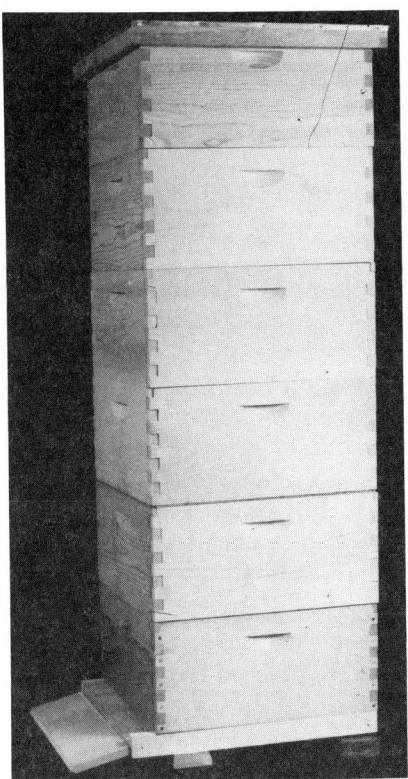

Figure 5-1, A Second Year Producing Hive and a First Year Producing Hive, Showing the Hive Parts. Left: The average second-year producing hive, consisting of two deep brood chambers and four deep honey supers. Right: The first-year hive, showing the parts that make up the hive. From the bottom: the bottom board, two brood chambers each with ten frames and foundation, the queen excluder, two honey supers each with ten frames and foundation, the inner cover, and the telescopic hive top. Notice the inclined board to the hive entrance and the slight forward tilt that prevents water from running into the hive.

was a mistake, for the bees did not get to feel any warmth on the nice winter days. As a result, they could not move to new stores of honey, so they starved to death surrounded by honey stores. "Killing the bees with kindness" is discussed in Unit 25, "Wintering The Hives."

The storage capacity of honey supers is approximately as follows.

Deep super: 3+ gallons of honey. Full super weighs about 80 lbs.
Medium super: 2+ gallons of honey. Full super weighs about 55 lbs.
Shallow Super: 1½ gallons of honey. Full super weighs about 45 lbs.

SELECTING HONEY SUPERS

The type of honey supers you select will depend on your strength and the type of honey you want to produce. The standard supers now in use in this country are of the same outside dimensions in width and length and differ only in depth. The standard width and length are 16¼ by 20 inches. Some manufacturers have changed the length recently to 19⅞ inches.

- *Deep supers* are 9⅝ inches deep, weigh about 80 pounds when well filled, and are used almost exclusively for brood chambers and for extracted honey.
- *Medium supers* are 6⅝ inches deep, weigh about 55 pounds when well filled, and are used for extracted honey, chunk honey, and comb honey.
- *Shallow supers* are 5¹¹⁄₁₆ inches deep, weigh about 45 pounds when well filled, and are used for extracted honey, chunk honey, and comb honey. Use this super if your are going to produce comb honey for sale in the frame or in plastic boxes.
- *Section comb supers* are 4¾ inches deep, weigh about 40 pounds when well filled, and are used to produce honey in the basswood section boxes. Kits are available to convert this super to the round plastic rings of comb honey. Four and one-half inch frames are available to convert this super to produce extracted honey, chunk honey, or comb honey.

TYPES OF HONEY YOU CAN PRODUCE
(See Figure 5-2.)

- *Extracted* or liquid honey is obtained by uncapping the combs of honey and using an extractor to spin the frames of honey, throwing the honey from the cells first from one side and then from the other. A radial extractor throws the honey from both sides of the comb during the spinning. The combs are not destroyed by extracting but can be used over and over, year after year. Most honey is extracted.
- *Chunk* honey is cut from the frames in strips, placed in a wide-mouthed jar, and the jar is then filled with extracted (liquid) honey.
- *Comb* honey is sold in the frame or cut from the frame and then cut to fit into a plastic box in which it is sold. Plastic boxes for this use are available from bee supply houses.
- *Section comb* honey is produced in basswood section boxes or plastic rings. These are now specialty honeys; in the past they were the standard product of most beekeepers. Section comb honey requires

ORDERING HIVE EQUIPMENT 21

Figure 5-2, Types of Honey You Can Produce. Left: A one-pound Queenline jar of extracted (liquid) honey. Right: A one-pound wide-mouth jar containing a segment of comb honey immersed in extracted honey, called chunk honey. The comb honey in the left background is comb honey in round plastic rings. The comb honey in the right background is comb from a shallow frame, cut to fit the plastic box in which it is sold.

considerably more management skill and time for a considerably smaller amount of honey produced by the hive than does extracted honey. If you plan to produce section comb honey, read some of the books available on the production of section comb honey.

WHERE TO BUY HIVE EQUIPMENT

It is best to buy your hive equipment from a bee supply house or a bee supply shop nearby rather than from the large, commercial, mail-order, department-type stores. Bee supply houses or shops sell only bee supplies and will have what you need—or can get it quickly. Mail or phone orders are filled and shipped within twenty-four to forty-eight hours. If there is a bee supply shop in your area, you can pick up and carry out the equipment you need as well as get advice and answers to most of your questions.

You can purchase hive equipment all year long. However, it may be best not to start a new hive any later than August 1, for even then you will have to provide sugar solution or frames of honey for winter stores.

WORDS OF CAUTION WHEN ORDERING WAX FOUNDATION

Many bee supply houses do not ship beeswax foundation to northern areas from about mid-October to about mid-March, because the foundation becomes very brittle in cold weather and may arrive in shattered pieces. It may be necessary to order foundation separately to arrive shortly before your package of bees arrives. If you have a bee supply shop nearby, you may be able to pick up the foundation; handle and store it carefully until needed in the spring. Usually no supers will be required for brood or honey so early that you cannot have foundation on hand when you need it.

Stored comb in the frames is every bit as brittle in cold weather; should the super be dropped or the stack of stored supers knocked over in freezing weather, it could be a disaster. If you must work with wax or foundation in cold weather, it is best to first bring it into the house where it can warm up to room temperature over several days.

ORDERING HIVE MATERIALS FOR EXISTING HIVES

- ☐ Take inventory of your hive materials as suggested in the first paragraph of this unit. You will be adding to your equipment only those things you will need.
- ☐ From the following lists, order supers of the size or sizes you intend to use. For example, if you find you need four deep supers, multiply each item in the deep super list by four and order that quantity of materials. The items shown in each list will make one complete super for honey storage when built up.
- ☐ Refer to the following key for clarification of references in the lists for ordering equipment.
    - (a) Wedge top bars, split or grooved bottom bars.
    - (b) Vertical wired/crimp wired or plastic reinforced foundation.
    - (c) Available only in multiples of 100, unless your bee supply shop sells in smaller lots.
    - (d) Thin surplus foundation is usually sold by the pound, about twenty-five sheets per pound.

Deep Supers for Honey Storage

If you are using only deep supers for honey storage, at the end of the second year your hive should have two brood chambers (deep supers) and four deep supers for honey storage. (See Figure 5-1.)

___(1) 9⅝" super
___(10) 9⅛" frames (a)
___(10) 8½" foundation (b)
___(20) support pins

Medium Supers for Honey Storage

At the end of the second year, your hive should have two brood chambers (deep supers) and six medium supers for honey storage. *Note:* If you are cutting comb for comb honey or chunk honey, you can probably run with five medium supers, because the bees will use up some honey in drawing new comb each year.

Extracted Honey

___(1) 6⅝" super
___(10) 6¼" frames (a)
___(10) 5⅝" foundation (b)
___(20) support pins

Comb or Chunk Honey

___(1) 6⅝" super
___(10) 6¼" frames (a)
___(10) 5⅝" cut comb foundation
___(20) support pins

Shallow Supers for Honey Storage

At the end of the second year, your hives should have two brood chambers (deep supers) and eight shallow supers. *Note:* If you are cutting comb for comb honey or chunk honey, you can probably run with seven shallow supers, because the bees will use up some honey in drawing new comb each year.

Extracted Honey

___(1) 5¹¹⁄₁₆" super
___(10) 5⅜" frames (a)
___(10) 4¾" foundation (b)
___(20) support pins

Comb or Chunk Honey

___(1) 5¹¹⁄₁₆" super
___(10) 5⅜" frames (a)
___(10) 4¾" cut comb foundation
___(20) support pins

Section Comb Supers for Honey Storage

At the end of the second year, your hive should have two brood chambers (deep supers) and six section comb supers. You will have fewer supers, because the bees will produce less honey in the section comb supers, and they will be drawing new comb each year.

Section Comb Honey in Basswood Boxes

___(1) section comb super, complete with section holders, etc.
___(28) split section boxes, 4¼" × 4¼" (c)
___(7) thin surplus foundation (d)

Plastic Rings of Comb Honey

___(1) section comb super, empty
___(1) kit to convert to plastic ring frames
___(9) thin surplus (d) or 4¼" cut comb foundation

To convert section comb supers to frames for honey storage, order the items below for each section comb super to be converted.

Extracted, Comb, or Chunk Honey

___(10) 4½" frames
___(10) 4¼" foundation

Supers of Mixed Sizes for Honey Storage

Suppose you have two brood chambers (deep supers) and two deep supers for extracted honey and would like to produce some chunk honey and cut comb honey (no section boxes). Order the following for the approximate honey storage capacity of four deep honey supers.

1 unit: Medium super with cut comb foundation = ⅔ of a deep super.
2 units: Shallow super with cut comb foundation = 1 deep super.

Because the combs will be cut and the bees will start on new foundation each year, the honey storage capacity here of approximately three and two-thirds deep supers would probably be enough—depending on the volume of floral sources in your area, of course.

ORDERING HIVE MATERIALS TO START NEW HIVES

For each new hive to be started, order the following. (See Figure 5-3.)

ORDERING HIVE EQUIPMENT 25

Figure 5-3, The First-Year Hive. If you start your hive as early in the spring as possible, you should expect to have a hive of this size by the end of the first year. Plan for this when ordering equipment. (See Figure 5-1.)

☐ For the brood area, order:
One standard ten-frame hive, which consists of the hive top, the inner cover, one deep super, ten frames, and the hive bottom. Foundation is not included. It must be ordered separately and is included in the following list. For the second brood chamber, order these items.

(1)   9⅝" deep super
(10) 9⅛" frames (a)
(20) 8½" foundation (b)
(40) support pins

☐ For honey storage, order the required number of units based on the size supers you want to use for each hive to be started.

Deep Supers

(2)   9⅝" deep supers
(20) 9⅛" frames (a)
(20) 8½" foundation (b)
(40) support pins

Medium Supers

(3)   6⅝" supers
(30) 6¼" frames (a)

(30) 5⅝" foundation (b)
    or cut comb foundation
(60) support pins

Shallow Supers

(4) 5¹¹⁄₁₆" or 5¾" supers
(40) 5⅜" frames (a)
(40) 4¾"foundation (b)
    or cut comb foundation
(80) support pins

Section Comb Supers

(4) section comb supers, complete
    with section holders, etc.
(120) section comb boxes, 4¼"
    split three sides (c)
(28) thin surplus foundation (d)

☐ Order the following accessories for each new hive.

(1) queen excluder
(1) bee escape
(1) entrance or Boardman feeder

ORDERING MISCELLANEOUS SUPPLIES FOR EXISTING HIVES

Remember to order supplies to restock your supers if you produced comb honey, chunk honey, or section comb honey last year. Also check your supplies of medicine on hand. Be sure you have enough for the existing hives plus any new hives you will add this year.

Pollen supplement is discussed later, but this first spring it is suggested that you order a pound for each of your hives to feed until you determine the early pollen availability in your area. Some areas have a dearth of early pollen, so a supplement should be fed to start brood rearing. Continue to feed until natural pollen is available in the field.

☐ If you have not done so, order enough foundation to replace the frames of comb cut from the shallow and medium supers last year.
☐ Order section boxes (twenty-eight per super) and thin surplus (seven sheets per super) for the section boxes removed last season.
☐ If you obtained the 6.4-ounce packet of Terramycin® (TM-25) last year, you should have enough. Otherwise, order the TM-25 in the 6.4-ounce packet, the smallest quantity available in most areas.

- ☐ The 0.5-gram vial of Fumadil-B® will medicate one hive for two years. Order if necessary.
- ☐ Order one pound of pre-mixed pollen supplement for each hive for spring feeding at the time you medicate. Some supply houses sell two pounds as the smallest quantity available.
- ☐ If you stored your supers sticky last fall, order one pound of Paradichlorobenzene (PDB) moth crystals to protect the supers from wax moth after the bees become active, and clean them up. After clean-up, remove from the hive and store the supers with frames in plastic bags with PDB. See Unit 22 for directions for storing clean supers.

YOU CAN SAVE MONEY IF YOU BUY IN QUANTITY

- ☐ Buy the super with frames as a unit, rather than buying a super and ten frames ordered separately.
- ☐ Buy in lots of five or ten units such items as complete ten-frame hives, supers, tops, bottoms, inner covers, queen excluders, etc. If you will need five in time, save by buying them at one time.
- ☐ Buy in lots of fifty or a hundred such items as frames and foundation. Foundation can be purchased in twenty-five pound lots or five pound lots for the cut comb or thin surplus.
- ☐ Buy section boxes in lots of 500 if you find you are going to continue to produce section comb honey.

UNIT **6**

Midwinter Hive Activities and Manipulations

In the North

- [ ] Keep the auxiliary exit/vent free of snow, ice, and frost.
- [ ] Heft the hive from the rear occasionally to keep a "feel" for the stores in the hive.
- [ ] Normally, *do not* attempt to open or enter the hive during the winter. However, if on one of those rare days in winter when the outside temperature reaches 65° F or more and you see activity at the auxiliary exit, you may unwrap the hive and have a *quick* look. Check for honey and pollen stores, brood (if any), and population strength before closing up the hive and rewrapping it securely. If you see activity at the entrance or auxiliary exit, that should be indication enough that your hive is doing well in its wintering, and it might be better not to go into the hive.
- [ ] Learn to recognize American Foulbrood Disease (AFB). Write to USDA, Washington, D.C., for Farmers' Bulletin #2255, "Identification and Control of Honey Bee Diseases." If at any time during the year you think you see symptoms of AFB, call the bee inspector or send off a sample for verification as directed in the bulletin.

In the South

- [ ] Keep the hive entrance free of leaves and debris. If leaves pile up at the entrance and become soaked by rain, the entrance may block completely, resulting in the death of the hive. In Unit 25, "Wintering the Hive," a vent/auxiliary exit is described that not only vents the hive (without boring holes in the supers) but also acts as an auxiliary exit—just in case.
- [ ] Heft the hive occasionally to keep a "feel" for the stores in the hive. In southern areas, bees fly on most days and probably use nearly as much honey stores over the winter period as in northern areas.

MIDWINTER HIVE ACTIVITIES AND MANIPULATIONS 29

The bees have the opportunity to gather limited stores from the field in their daily foraging. However, much of these stores are used up in the limited brood rearing that could go on almost all winter long.

☐ Any day on which the temperature is above 65° F, the hive may be inspected for honey and pollen stores, brood rearing, and population strength. During the winter, keep these inspections of a short duration. The hive can be fed almost any time in winter if the hive weight indicates the need for additional stores. Nosema medication can be fed during the winter at two-week intervals, insuring that the medicine is used for winter food and for the control of Nosema Disease. See Unit 23, "Fall Medication and Feeding of the Hive."

☐ Learn to recognize American Foulbrood Disease (AFB). Write to USDA, Washington, D.C., for a copy of Farmers' Bulletin #2255, "Identification and Control of Honey Bee Diseases." If at any time during the year you think you see symptoms of AFB, call your bee inspector or send a sample off for verification as directed in the bulletin.

PART **III**

EARLY SPRING ACTIVITIES AND MANIPULATIONS

The next five units cover the earliest activities and manipulations of the hive. These activities will prepare your hive to make the best showing during the honey flow.

- **Unit 7** A PREVIEW OF EARLY SPRING ACTIVITIES AND MANIPULATIONS
- **Unit 8** ABOUT BEE DISEASES
- **Unit 9** A REVIEW OF SOME IMPORTANT ACTIVITIES AND MANIPULATIONS
- **Unit 10** THE PREMEDICATION CHECK
- **Unit 11** EARLIEST MANIPULATIONS OF THE HIVE

Unit 7

A Preview of Early Spring Activities and Manipulations

In the North: As the weather moderates with the coming of spring and the temperature reaches 60° F or more for a few hours of each day, you should begin to see activity at the hive entrance or at the auxiliary exit. It is possible that the hive entrance could be blocked by dead bees. The activity at the hive is reorientation following the winter clustering, housecleaning of winter debris and dead bees, and the search for water to dilute honey to a nectar consistency for feeding the larval stage of the brood. Brood will be produced in increasing numbers as the days become warmer.

Activity at the hive entrance is not a guarantee that your hive has made it through the long, cold winter. Your hive might be dead, and the activity you see could be bees from another hive robbing out the stores from your hive in a leisurely fashion for their own use. This can also happen to very weak hives at this season. If an early pollen source is available in your area and you see bees returning to the hive with pollen on their back legs, you can be quite sure your hive has made it through the winter.

If there is no activity at the hive entrance after about ten days of pleasant weather of 60° F or more for part of each day, you should determine whether your hive has died over the winter or is in a very weakened condition—possibly even on the verge of starvation. The hive should have a premedication check, which is described in Unit 10.

Once the weather is temperate, clean and set up your watering system, because the bees require extra water in the spring for "watering down" honey to feed the earliest brood. Now is the best time to start training your bees to a constant source of water near their hive location. If you have a year-round stream, creek, or spring within half a mile of the hive, the bees will use that source in preference to your system or a neighbor's pool. Check occasionally to see that the natural water source has not dried up. If the water source begins to dry up, get your

34 KEEPING BEES

watering system back in use. Better still, keep your system set up and working all the time. (See Figure 7-1.)

During the spring, the active hives that made it through the winter are discovering the hives that died out over winter or those too weak to keep guards at the entrance. It is important at this time to medicate

Figure 7-1, Watering Systems. Top left: a bucket with floating wooden blocks. Top right: A pan with a towel covering two floating two-by-fours. Bottom: A ceramic pot filled with old macrame plant hangers (old rope or binder twine would also work.)

your hives to control the spread of the foulbrood diseases. If a nearby hive dies as a result of AFB, the disease can be spread by your bees robbing out the diseased honey. Spring medication is covered in Unit 11.

Medicating to control Nosema Disease gives you an opportunity to feed your hive and at the same time stimulate brood rearing. The Nosema medicine is fed in a sugar solution that provides extra stores to keep the hive going (if it needs help) until early nectar is available in the field.

Pollen supplement, available early in the spring, may stimulate brood rearing in areas where an early pollen source is not readily available. Feed a pollen supplement the first spring to determine if early pollen is available in your area. If early pollen is available, the bees will take very little of the pollen supplement patties, and in that way they will tell you that they do not require the pollen supplement. If after a year or two the bees have not taken the pollen supplement early in the season, discontinue feeding it. Preparing pollen supplement and dispensing it is covered in Unit 11.

Brood rearing in a two-story hive—a hive with two brood chambers—usually begins in the upper brood chamber. Remember, the brood must be kept at a temperature of 92-94° F to develop normally. Pollen stores are as important as nectar/honey in brood rearing; brood rearing cannot take place if either is missing.

SOME REASONS YOUR BEES MIGHT NOT HAVE MADE IT THROUGH THE WINTER

- *Winter kill:* An extremely cold, windy winter with no day for a period of six to eight weeks sunny or calm enough for the hive to warm sufficiently for the bees to break cluster and move to new stores results in winter kill. The bees starve to death while surrounded by stores of honey.
- *Insufficient honey stores:* If there are not enough stores to last through the winter, the bees starve after using all available stores.
- *An old queen or the loss of the queen:* If young bees are not reared in the months before the bees cluster for winter, the old bees may die off during the winter, so reducing the population that the cluster cannot survive. If the queen is lost in the fall, the hive may not have enough young bees going into winter and too many old bees may die during the winter, so the hive may not survive.
- *Disease in the hive:* With disease of any sort in the hive, but particularly AFB, fewer larvae live. As a result, there are fewer young bees for the winter cluster, and the hive stands a greater chance of

dying during the winter. These dead hives are the ones that can bring the diseases to your other hives. Nosema Disease causes the adult bees, young and old, to die more quickly during the winter.

- *Damp conditions:* Dampness in the hive results from lack of ventilation. The bees can tolerate a lot of cold but cannot live with *damp* and cold. If the hive is not vented to some small degree, the moisture vapors created by the metabolic processes of the bees in the cluster can produce water condensation, frost, and icicles that greatly increase the death rate of the bees in the cluster. The hive may not survive.

In the South: In the southern areas the bees may fly almost every day in winter, except when kept in the hive by rain and high winds. If your bees are confined to the hive for four weeks or more by continual cold weather, consider youself *In the North* for beekeeping purposes.

Diminished activity at the hive entrance during the winter season means you should plan to have a quick look into the hive on the first pleasant, calm day when the temperature is at least 60-65°F. If your hive seems weak, close the hive entrance to one inch. Your hive could be weak or could have died out over the winter and be leisurely robbed out by another hive in the area or a hive in your own apiary. Always watch for pollen coming into the hive, and remember that robber bees do not bring in or steal pollen. Pollen coming into the hive is always a good sign.

Heft the hive from the rear or side occasionally during the winter season. If the hive seems to lighten over time, it is a good idea to feed the hive slowly until nectar becomes more readily available in the field. Feeding slowly means feeding a quart every three or four days rather than feeding one quart after another as fast as the bees will take it.

Keep a watering source available for the bees the year round. If there is a year-round natural water source of some sort within a half mile, the bees will find it and not bother the neighbors' pools, bird baths, or fish ponds. Keep a watering source always available if the natural source is apt to dry up. (See Figure 7-1.)

Your hive, if active and strong, may rob out weak and dead hives in the area. If the weak or dead hive is diseased, your bees could bring the diseases back to your hive. Under these conditions, it is very important to medicate early in the brood-rearing season to control the foulbrood diseases. In the South, AFB/EFB medication should be started about ten weeks before the honey flow begins. In the far South, start medication about January 15. In the northern limits of the southern areas, start medication about February 15. In the San Jose, California, area, the author starts medicating on February 1, followed by medications on February 10 and 20. Spring medication will be covered in Unit 11.

A PREVIEW OF EARLY SPRING ACTIVITIES AND MANIPULATIONS

Wax moth is a problem for the beekeeper in the South about ten months of the year. Only during the winter period, when the days are short and the days and nights are relatively cool, are the very weak or dead hives not destroyed quickly by the wax moth. Apparently the metamorphic cycle of the wax moth is slowed down at that time of the year. After extracting, empty supers should be cleaned up by the bees and stored with Paradichlorobenzene (PDB) moth crystals through the winter and until needed during the honey flow. Napthalene moth crystals are not at all effective in controlling the wax moth. Wax moth is covered in Unit 27.

In most areas of the South, pollen supplement is not required, but plan to feed pollen patties along with the medications for the first year or two until you know if adequate early pollen sources are available in your area. Remember that pollen is as important as nectar/honey in the brood rearing process. If either pollen or nectar/honey is unavailable, brood rearing cannot proceed. Pollen traps are available and enable the beekeeper to steal pollen (one day out of four during the season when it is abundant) to use for feeding back natural pollen to the bees in the spring, or as an organic natural high-protein supplement in the human diet. Pollen supplement is discussed in Unit 11. (See Figures 34-1 and 34-2.)

In the South, the reasons your bees might not make it through the winter are the same as those in the north, with the exception of winter kill.

UNIT 8

About Bee Diseases

LEARN TO RECOGNIZE THE FOULBROOD DISEASES (AFB/EFB)

A good, responsible beekeeper should learn to recognize the foulbrood diseases, particularly American Foulbrood Disease (AFB). If you have not already done so, immediately sit down and write to the USDA, Washington, D.C., and request a copy of Farmers' Bulletin #2255, "Identification and Control of Honey Bee Diseases."

Learn to recognize the symptoms of the foulbrood diseases. Check for symptoms each time you inspect the brood chambers of your hives. If you suspect that you may have a foulbrood disease, either contact your county bee inspector or send a sample of the suspected material to the address shown in Farmers' Bulletin #2255. Your bee inspector may be able to give you the address of the state laboratory that can diagnose suspected AFB samples. Eventually you will be able to determine for yourself whether or not your hives have the disease.

Steps for preparing and dispensing the needed medications are covered in detail in Unit 11.

AMERICAN FOULBROOD (AFB)

American Foulbrood is caused by a spore-forming germ, *Bacillus larvae*. Because of the germ spores, AFB is the most destructive of the bee diseases. The spores can be destroyed only by burning the frames and combs and scorching the hive body, top, and bottom. AFB is a disease that affects only the brood of the honey bee. Adult bees and humans are not affected in any way by ingesting the spores of *Bacillus larvae*. AFB is spread to your hive in the following ways.

1. Your healthy bees rob a hive that died of AFB or one that has been weakened by the disease. Spores of the disease are brought back to

the hive in the contaminated honey stores from the diseased hive or on the hairs of the bees' bodies; the spores fall off and contaminate the honey/nectar and pollen stores in your hive.
2. The beekeeper buys old, used equipment or established hives without inspecting for AFB or without knowing the symptoms of AFB. AFB spores are suspected of remaining viable for 100 years, and it is known positively that the spores are viable for over forty years. Buying old equipment not used for many years could give your bees AFB from spores hidden away for all those years.

AFB can be spread around your apiary and your neighborhood by: (a) bees robbing out diseased hives; (b) using old equipment from an unknown source or bringing in diseased hives; (c) feeding diseased honey to the hives or permitting the bees to clean up extracting equipment that may have extracted diseased honey; or (d) mixing supers from diseased hives and healthy hives, thereby transmitting the disease to the healthy hives.

Do not restock a hive that has died of a foulbrood disease without cleaning up the hive equipment as described in Farmers' Bulletin #2255. Most states require that AFB-contaminated hives be burned.

Controlling American Foulbrood (AFB)

Terramycin®, which contains the antibiotic Oxytetracycline, is effective in controlling American Foulbrood. It is not, however, a cure for the disease once the disease has started in the hive.

A conscientiously applied medication routine, year after year, can do much to prevent the disease from getting started in your hives. The medication works as follows: When the spores of AFB are fed to the larvae in the honey/nectar under proper conditions of warmth and moisture, the bacillus bursts from its spore casing as a living germ. The Oxytetracycline in the honey/nectar that is being fed the larvae kills or retards the germ, and the larvae live. If most of the larvae live in the early spring (or in the fall), the population increases steadily, and the increased work force rapidly cleans out and carries away dead larvae and pupae before the bodies rot and the millions of bacillus germs revert to the spore form.

The bees are most apt to rob out dead and weak hives that may have AFB at those times of the year when the least nectar is available in the field; namely, early spring and after the honey flow in the fall. These are the best times to medicate the hives to control any spores in the diseased honey from bursting into life and infecting the larvae and the hive. If the medication is fed to the hive at the time spores of AFB are being brought into the hive, the medication will be stored in the brood chambers and fed to the larvae along with any spores of AFB that are in the diseased honey.

It should also be noted that at the time the spores of AFB are being brought into the hive, the hive population is at its lowest. If the AFB disease begins to kill off the larvae and pupae, there are not as many bees to clean out the dead brood. If you medicate and keep most of the larvae/pupae alive that would otherwise die, the population will be considerably larger by the time you stop medicating, and any dead larvae/pupae will be removed before the bodies decay.

When medicating, the medicine should be spread over the largest area of brood and honey/nectar stores, and it should be made available over an extended period of time when the diseased honey would be brought into the hive. Stretching the medicine with more sugar and medicating several times rather than just once will let you meet the requirements above.

EUROPEAN FOULBROOD (EFB)

European Foulbrood is caused by the germ *Streptococcus pluton*, which does not spore. As a result, EFB is easier to control than AFB. Introducing a new young queen and medicating with Terramycin® will bring the disease under control.

EFB is spread in the same manner as AFB, primarily by the robbing of contaminated honey from diseased weak or dead hives; the exchange of sticky, contaminated supers with those of your healthy hives; or permitting bees to clean up contaminated honey or honey equipment.

Controlling European Foulbrood (EFB)

Terramycin®, which contains the antibiotic Oxytetracycline, is effective in controlling European Foulbrood. The same conscientiously applied medication program used to control AFB will also control EFB. It is not necessary to burn EFB-contaminated hive equipment. The disease usually can be controlled by re-queening and routinely applying Terramycin®.

COMPARISON CHARACTERISTICS OF THE FOULBROOD DISEASES

Table 8-1 is a guide for comparing characteristics or symptoms of the American Foulbrood and European Foulbrood diseases. Do obtain Farmers' Bulletin #2255 for photographic details of the honeybee diseases.

TABLE 8-1   Comparison Characteristics of AFB/EFB *

CHARACTERISTICS TO OBSERVE	AMERICAN FOULBROOD	EUROPEAN FOULBROOD
Appearance of brood comb.	Sealed brood. Discolored, sunken, or punctured cappings.	Unsealed brood. Some sealed brood in advanced cases with discolored, sunken, or punctured cappings.
Age of dead brood.	Usually older sealed larvae or young pupae.	Usually young unsealed larvae; occasionally older sealed larvae.
Color of dead brood.	Dull white, becoming light brown, coffee brown to dark brown, or almost black.	Dull white, becoming yellowish white to brown, dark brown or almost black.
Consistency of dead brood.	Soft, becoming sticky to ropy.	Watery to pasty; rarely sticky to ropy.
Odor of dead brood.	Slight to pronounced glue odor to glue-pot odor.	Slightly to penetratingly sour.
Scale characteristics.	Uniformly lies flat on lower side of cell. Adheres tightly to cell wall. Fine, threadlike tongue of dead pupae adheres to roof of cell. Head lies flat.	Usually twisted in cell. Does not adhere tightly to cell wall. Rubbery.

*U. S. Department of Agriculture. "Identification and Control of Honey Bee Diseases." Farmers' Bulletin #2255. Washington: G.P.O., 1978, 18pp.

NOSEMA DISEASE

Nosema Disease is caused by the spore-forming protazoan, *Nosema apis*. The spores of the disease are ingested by the adult bees; they germinate and multiply in the midgut of the bee.

Nosema Disease decreases the effective life of the adult workers, resulting in a decrease in brood rearing and honey harvest. The disease can cause succession of the queen in the hive. Nosema Disease is most noticeable in the spring when the dysenteric condition tends to be present in the hive. The disease is spread within the hive by the infected bees defecating on the brood and honey combs, which contaminates the honey and pollen stores. Healthy bees become infected as they clean up the combs, and they spread the spores around the hive on the hairs of their bodies. Bees robbing a hive infected with Nosema Disease will carry the spores back to their hives in the infected honey and on the hairs of their bodies. The spores are placed or dropped in the honey or pollen stores.

Controlling Nosema Disease

Fumadil-B®, containing the antibiotic *Fumagillan*, is used for the control and prevention of Nosema Disease. To decontaminate soiled bee equipment, follow the directions in Farmers' Bulletin #2255.

DIAGNOSIS OF BEE DISEASES

If you own a microscope or have access to one, you can diagnose bee diseases by obtaining from the USDA in Washington, D.C., the Agricultural Research Service Bulletin, ARS-NE-87 "Diagnosis of Honey Bee Diseases, Parasites and Pests." If you do not have access to a microscope, send samples to the address given in Farmers' Bulletin #2255 for diagnosis.

UNIT 9

A Review of Some Important Activities and Manipulations

Before going to the hive for the first time in the spring, review the activities and manipulations that you will almost always perform during any hive manipulation. These are: (1) dressing properly, (2) lighting the smoker, (3) entering the hive, and (4) inspecting the brood chamber. Several other activities will be reviewed that you can hope will be performed only a few times during the year. These are: (1) locating the hive, (2) moving the hive, and (3) removing a bee stinger.

EXPLANATION OF STEP-BY-STEP PROCEDURES

In this handbook the various steps of any hive manipulation are indicated by the following notation:

☐ Remember this.

Each notation is a single step in the manipulation. It is essential for your safety and that of your bees to perform the steps given in the initial description of that procedure before going on to the next step in the manipulation.

For example, ■ **Dress properly,** ■ **Light the smoker,** and ■ **Enter the hive** will appear as steps in most manipulations. When such notations appear, perform *all* the steps as described in this and other sections. If you need to refresh your memory at a later time, refer again to the full step-by-step instructions given in the initial description of the process.

MATERIALS FOR DRESSING PROPERLY TO WORK THE BEES

By wearing the following items of apparel the beekeeper will receive few stings during the beekeeping year. (See Figure 9-1.)

44 KEEPING BEES

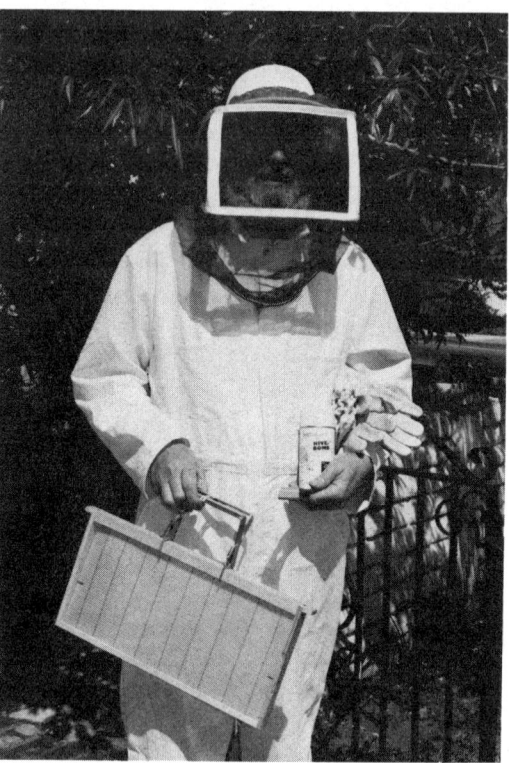

Figure 9-1, A Well-dressed Beekeeper. The author in new coveralls with a zippered veil, holding leather gloves, smoke bomb, and hive tool in one hand, and a frame grip holding a frame of vertical wired foundation in the other. Not shown are the boot bands holding the coverall legs securely against the boots.

(1) coverall and (1) veil or (1) coverall with attached zippered veil
(1) helmet
(1) pair of leather beekeeping gloves (Get rid of the canvas gloves.)
(1) pair of boot bands
(1) sweat band to keep perspiration out of the eyes
(1) pair of boots, high shoes, galoshes, or leggings

- *Coveralls* should be white or of light pastel shades. Avoid dark colors or red. The coveralls should be of cotton, nylon, or cotton/dacron material and should have a full-length zipper with a button or snap at the neck. Pass-through pockets (allowing access to trouser pockets) should be sewn shut. The coveralls should be a size four inches larger than your chest size, and you should try them on to be certain of a loose but good fit. Coveralls should not fit snugly; a more or less "sloppy" fit is best, so the coveralls can go over a jacket in cooler weather. Bee supply houses and shops carry coveralls with an attached zippered veil or with no veil attached. The coverall with the zippered veil is probably more bee-proof than a coverall with a drawstring veil.

However, each type of veil arrangement has its advantages and disadvantages.
- The *veil* can be a round or a folding wire screen veil. Both types are available with either a drawstring or a separating zipper, which can attach to coveralls. There are many styles and types of veils, so choose the one that suits you. Most veils fit a hat with a brim no wider than three inches.
- The ventilated *helmet* with an adjustable headband, called a pith or jungle helmet, is the best headgear. A straw hat is also good. If the brim of the straw hat is more than three inches wide, cut it to the proper width and sew or glue cloth tape to the edges to prevent fraying.
- *Leather beekeeping gloves* are sting-proof, whereas canvas and plastic-coated gloves are not. Leather beekeeping gloves are recommended and are a good investment.
- *Boots*, high shoes, or galoshes should be worn so the pantlegs can be tucked into or fastened around the boots with boot bands. This keeps the bees from stinging the ankles or crawling up inside the pantlegs.
- *Boot bands* can be purchased or made at home. They are used to hold the pantlegs of the coverall securely to the boots, keeping the bees from stinging the ankles or crawling up inside the pantlegs.
- A *sweat band* is useful if you tend to perspire heavily. Put on a sweat band or tie a bandana around your forehead before putting on the helmet and veil.

■ DRESS PROPERLY

☐ Put the coverall on over whatever you are wearing. Turn up the collar, zip fully, and snap or button at the neck.

☐ Put on your boots, high shoes, or galoshes. Tuck the pantlegs into the boots and tie the boots securely or use boot bands, elastic bands, bicycle clips, or leggings to hold the coverall legs securely around the boots.

☐ Slip the veil and helmet over your head; the lower part of the veil goes over your collar. If a drawstring veil is used, close the drawstring to about four inches below the neck and tie the veil in back—although it is always more secure if you can cross the ties in back and tie off in front. Fluff out the netting of the lower portion of the veil to keep it and the bees from your neck.

If your coveralls have the attached zippered veil, slip the veil and helmet over your head and zip the veil to the coveralls. Fluff out the netting on the lower portion of the veil to keep it away from your neck to prevent any bees that land on the netting from stinging. (See Figure 9-1.)

- [ ] Before putting on the gloves, ■ **Light the smoker** or have the hive bomb handy (this will be discussed fully later). When the smoker is lighted, pull on the gloves and pull the gauntlets up over the coverall sleeves as far as they will reach.
- [ ] Pump the bellows on the smoker once or twice, pick up your hive tool, frame grip, bee brush, and whatever else you need for the manipulation—and you are ready to work the bees.

Note: Do not wear any woolen outergarments or a felt hat when working the bees. The bees sting materials of animal origin much more quickly than materials of vegetable origin (cotton or straw), chemical fabrics, rubber, or properly tanned leather. The scent of venom from stings in the flesh or clothing excites other bees to sting.

If you have no coveralls, follow the steps below for best protection from bee stings.

- [ ] Step into a second pair of pants (no wool, please). A sloppy-fitting pair of cotton trousers is fine.
- [ ] Put on your boots, high shoes, or galoshes. Fasten the pantlegs closed by tucking them in the boots or by using boot bands, elastic bands, leggings, etc.
- [ ] Put on a jacket (again, no wool) of cotton, nylon, cotton/dacron, or leather. Tie the jacket snugly at the hips to keep the bees from getting under the jacket. Zip the jacket to the neck and lock the zipper, then turn up the collar.
- [ ] Slip the veil and helmet over your head, and draw the string of the veil up to about four inches from your neck. Pull snugly and tie the strings either in back or, if the strings are long enough, in front. Fluff out the netting of the lower portion of the veil to keep it and the bees from your neck.
- [ ] Before putting on the gloves, ■ **Light the smoker** or have the hive bomb handy. Pull on the gloves and pull the gauntlets up as high as they will reach over the jacket sleeves.
- [ ] You are now ready to work your bees. Remember to pump the bellows of the smoker at one- or two-minute intervals to keep it going.

SMOKING THE HIVE

The smoker is the beekeeper's best friend. (See Figure 9-2.) A gentle puff or two of smoke at the entrance and under the hive cover sends the guard bees into the hive, where they begin to gorge themselves with honey. With the guard bees out of the way, you can carefully enter the hive and do whatever is necessary.

A REVIEW OF SOME IMPORTANT ACTIVITIES AND MANIPULATIONS 47

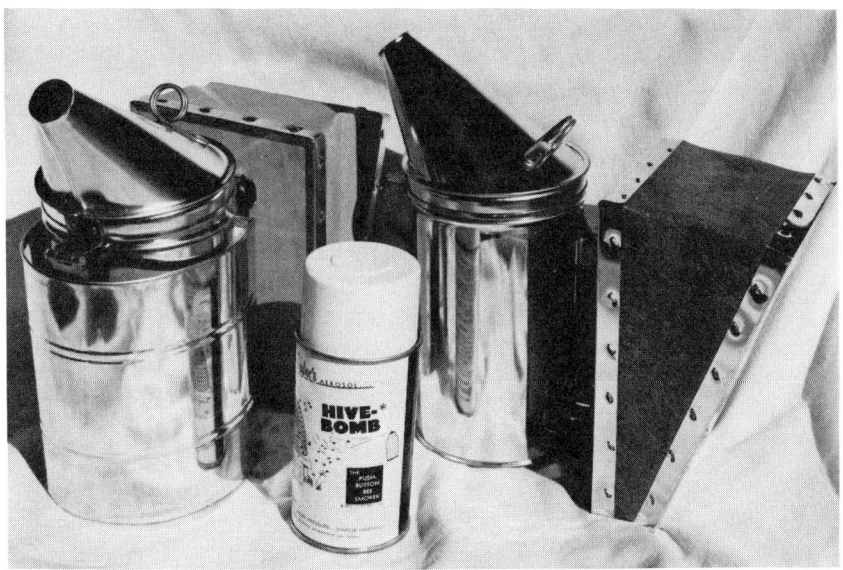

Figure 9-2, Smokers and Smoke Bomb. The smoker on the left is a shielded smoker so the beekeeper--and his little helpers--cannot burn themselves. The smoker on the right is the same smoker without the shield. The bellows on most smokers can be replaced. The can in the center is a smoke alternative: a push-button aerosol containing a smoke scented material.

You *should not* try to enter the hive without smoking it, nor should you try to work the hive without dressing properly. It is true that during the honey flow the bees can be very easy to work—perhaps without smoke or veil—but tomorrow may be altogether different, and you could be stung many times for your foolhardiness. Mother Nature is a good, convincing teacher. If advice from this beekeeper does not convince you, one day the bees themselves will!

A Smoker Alternative

Available at most bee supply houses and shops is a pressurized can whose contents have the scent of smoke that, in the author's view, is as effective as smoke from burning materials in a smoker. The aerosol smoker, which is called the hive bomb in this volume, is cooler, needs no lighting or pumping to keep it going, and is always ready. The aerosol goes out only once—and that is when the contents of the can are used up. The aerosol smoker should last most of a beekeeping season for one or two hives. *Note:* If your hives are in an area of fire danger, you might want to have and use the aerosol smoker to make sure you do not start a fire, which could be done accidentally while lighting or using the conventional smoker.

The Smoker

Although other sizes are available, the four-inch diameter by seven-inch deep smoker is the most satisfactory size for the hobbyist beekeeper. The smoker should last for years with normal care, but remember that when in use, the smoker contains burning materials and can cause burns. A lighted smoker *must not* be carried in a truck or car, because it could cause a fire.

Practice using your smoker on an empty hive. Learn how long it will stay lighted after it has been started and when given an occasional pumping of the bellows.

Smoker Fuels

As fuel for your smoker, you want a material that will light easily; smolder readily; and give off a cool, dense smoke. Probably the best smoker fuels are burlap or pine needles. Burlap is easiest to light but burns faster than most other materials. Pine needles start a bit harder but burn for a good while and create a cool, dense, white smoke.

Smoker fuels like burlap, cotton cloth, cardboard, etc., can be made up into rolls about 5 inches long and 3½ inches in girth and tied with string. Bulk materials such as pine needles, straw, grass, cobs, wood shavings, etc., are added by the handful when necessary. *Remember*, whatever you use for smoker fuel, you must pump the bellows every minute or two to keep the materials inside smoldering, or the smoker will go out—just when you need it most. *Do not* burn in the smoker any material of animal origin, plastic, or synthetics.

■ LIGHT THE SMOKER

1. If you are using burlap rolls, hold a lighted match under the end of the roll of burlap to start it burning. Turn the roll as burning continues. When the burlap is burning well on the end, drop it into the smoker can, close the lid, and pump the bellows a few times, and you are ready to put on your gloves and work the bees.

2. If you are using rolls of some material other than burlap, follow the steps below for lighting the smoker.

- [ ] Take one-half of a single sheet of newspaper, crumple, light with a match, and push it down into the smoker can with the hive tool. Pump the bellows several times to keep it burning.
- [ ] Place a roll of the smoker fuel to be used in the smoker tank so that it sticks out about one inch. Pump the bellows briskly to ignite the roll and get it burning well. When the roll of material is burning well (and this may take some practice to know when it is burning well enough not to go out after you close the smoker lid), push the roll down into the can with the hive tool, close the smoker lid, and pump the bellows several times. Put on your gloves and work the bees.

3. If you are using bulk materials, follow these steps.
   - [ ] Take one-half of a single sheet of newspaper, crumple, light with a match, and push it down into the smoker can with the hive tool. Pump the bellows several times to keep it burning.
   - [ ] Add a handful of bulk material onto the burning paper and pump the bellows to get it burning. Add another handful of bulk material and pump the bellows to get the materials burning well. Stuff the can loosely full of the bulk material and pump the bellows vigorously several times to get the burning going well. Close the cover of the smoker, pump the bellows a few times, put on your gloves and work the bees.

■ ENTER THE HIVE

Each time you enter the hive to work the bees you should perform the following steps, which in time you will memorize.

Equipment needed:

- [ ] Hive tool
- [ ] Smoker or hive bomb
- [ ] Frame grip

Step-by-step directions:

■ **Dress properly.**
■ **Light the smoker** or have hive bomb handy. Smoke the hive entrance.
  Note: Later in this handbook you will see the following comment after ■ **Enter the hive:** "You need not smoke the hive entrance." Most experienced beekeepers do not smoke the hive entrance when entering the hive. They smoke under the top cover or between the supers they are going to manipulate. They also work the hive from behind or from the side.

☐ A. *If you have an inner cover under the top cover*, follow these steps; otherwise, go to B.
☐ Lift the hive top and puff smoke in the oval hole of the inner cover. Remove the hive top and invert it on the ground nearby.
☐ With the hive tool, lift the inner cover from behind or from the side about one inch and puff in smoke over the frame tops, then lower the inner cover.
☐ Wait one minute for the smoke to take effect. Gently remove the inner cover and pump several puffs of smoke over the frame tops. The inner cover can stand nearby or can be laid near the front of the hive.
☐ B. *If you have no inner cover under the top cover*, follow these steps.
☐ With the hive tool, lift the hive cover about one inch, or until you can puff smoke under the hive cover. Lower the hive top.
☐ Wait one minute for the smoke to take effect. Gently remove the hive top and puff smoke over the frame tops. The cover can be inverted on the ground nearby or leaned against something.

You are now ready to medicate, feed pollen supplement, or (later) to inspect the brood chambers, add honey supers, etc., as directed for those manipulations.

LOCATING THE HIVE

The ease of entering the hive depends a lot on its location, so the following is a discussion about some aspects of locating the hive. You should have your hive where you can pull up a lawn chair and spend some time relaxing while watching the bees. It is even all right to talk to the bees.

The hive should be located so that it can be worked from behind and at least one side. You should seldom, if ever, work the hive from the front. Working the hive from the front will interfere with hive traffic, and you stand a greater chance of being stung. You should be able to stand and move in the area near the hive. You should not have to crouch down or lean over or be squeezed between things to get to the hive.

The bees can be encouraged to fly up and over the heads of pedestrian traffic by facing the hive toward and about four feet from a fence, or by placing a barrier in front of the hive, such as a short fence or a vine-covered trellis. Place the hive about four feet from the obstruction, facing it.

The hive should be set on blocks, bricks, or two-by-fours to keep the bottom dry. It should be level from side to side but should tilt forward slightly (a ¾" scrap of wood under the rear of the hive will do nicely). It is a good idea to place a sloping board from the ground to the entrance of the hive or, if the hive is on a bench, extend the bottom board with a scrap piece of plywood. The reason for such a ramp is this: In the early spring and fall, the bees returning late in the day are cold, tired, and heavily laden. If they land short, in most cases they die on the ground in front of the hive, unable to fly the few inches up to the hive entrance.

There are ideal locations for the bees, but most of us do not have them in our backyards, so do not worry whether your bees get too much or too little sun. Place the bees where it is the most convenient for you, where they will not be flying over congested areas, and where you can work them easily.

MOVING A HIVE OF BEES

Once the bees are in the hive, consider it a permanent location, and do not casually move the bees from one part of your yard to another. Bees orient visually to their location and, once in the air, find their way home again not so much from the appearance of the front of their hive as by nearby landmarks. Bees can reorient to a very minor or a very major change in location, but a change of between ten feet and half a mile only confuses them.

In the North: The best time to move the hives in the North is in the late fall before wrapping the hives for winter or early in the spring before the hives become active. After closing up the front you should fasten the brood chambers together with hive staples before moving the hive. You can move the hive any distance in the yard at this time of the year, because the bees will be reorienting when they once more become active in the spring. (See Figure 9–3.)

If you move the hive after the bees become active and have oriented to the hive location, you must follow the procedure given below for *In the South*. When oriented to a hive location, bees can be moved only a few feet or more than about two miles or they become lost. A move of ten feet will confuse the bees; they may not be able to locate the hive and may die during the cold nights out in the open. Bees apparently freeze to death at about 43°F if they are alone in the open or if not enough bees are clustered together for all to keep warm.

In the South: If your bees are active most days, you must move the bees no more than about two feet in any direction each day. You should not rotate the hive to change the entry direction more than about 45

Figure 9-3, A Hive Ready for Moving. The hive entrance is closed with a roll of plastic or nylon screening, hive staples attach the bottom board to the brood chamber, and the top is taped to the brood chamber so it cannot be blown off in transit.

degrees at any one time each day. Continue the short moves or small rotations until the hive is in its new location or facing in the proper direction.

The best way to move bees almost any distance (anything over fifteen feet) is to close the entry with a roll of screen cut to size. Using hive staples, fasten the bottom to the brood chamber and the lower brood chamber to the upper brood chamber. (See Figure 9-3.) Always close in the bees before pounding in the hive staples; this should be done about dark. When the hive is closed and stapled, load the hive and move it two miles or more away for a temporary stay of two weeks. When you unload the bees at their new location, pull out the screen. The following day the bees will reorient. Wait two weeks. After dark, close the hive entry with the screen roll and move it back to your yard into the desired location, pull out the screen, and the bees will reorient again the next day, having forgotten the orientation to the original location in the yard. Follow the steps below for moving the hive when moving it two miles or more.

Equipment needed:

- [ ] Hive staples
- [ ] Hammer
- [ ] Screen roll to close hive entrance

Step-by-step directions:
- ■ **Dress properly.**
- ■ **Light the smoker** or have hive bomb handy.
- ☐ After dark, close the hive entrance with a roll of screen. Once the hive entrance is closed, pound a hive staple at an angle near each corner joining the bottom board to the lower brood chamber and joining the lower brood chamber to the upper brood chamber.
- ☐ Tape the top to the upper brood chamber so it will not blow off in transit. Load and move the hive directly at that time, or load the hive and move it early in the morning. Move the hive at least two miles away. Do not keep the bees confined too long in hot weather (or in any weather, as far as that is concerned).
- ☐ When the bees are unloaded in the new, temporary location, pull out the roll of screen. The bees will orient to the new location during the next day and none will get lost.
- ☐ *Wait two weeks.*
- ■ **Dress properly.**
- ■ **Light the smoker** or have the hive bomb handy.
- ☐ After dark, place the screen roll in the hive entrance. Load the hive at once, and return it to the new location in your yard. Or load the hive at once, and return it to the new location early in the morning.
- ☐ With the hive in its new location, pull out the screen roll. The bees will orient to the new location in your yard during the day.

Note: If the bees are hanging out on the hive entrance when you are ready to put in the screen roll (usually in warmer weather), give them a puff or two of smoke and wait about five minutes. Most of the bees will go inside to find honey. In some cases, during quite warm weather, you may have to repeat the smoking several times. When most of the bees are inside, you can push in the screen and brush away the few bees remaining on the outside. Wear your gloves when loading and unloading the hive of bees.

REMOVING A BEE STINGER

If you are stung on the hands, face, or any part of the body where the bee does not sting through clothing, a stinger will be left in the skin. Remove the stinger by scraping it off at skin level with a knife blade, the hive tool, a fingernail, or any flat surface. *Do not* pick the stinger out with tweezers or with the fingers, for if you do, you squeeze the poison sack that is attached to the stinger, squeezing the full load of venom into your body. Remove the stinger as soon as possible, for it tends to pull itself deeper into the skin and pump more of the venom into your body by involuntary muscles that continue to work after the

54 KEEPING BEES

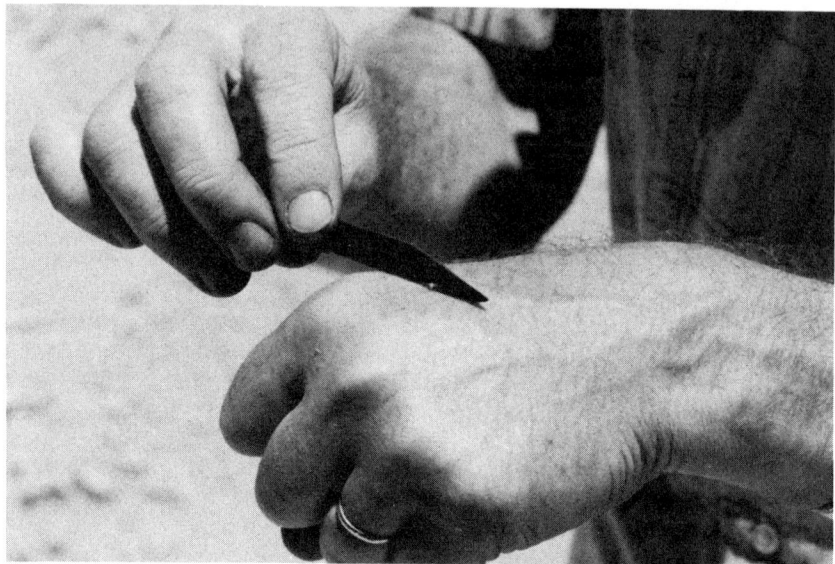

Figure 9-4, Removing a Bee Stinger. The stinger should be removed by dragging it out with a pen knife, hive tool, fingernail, or other flat, narrow surface (such as a credit card). If you grab the stinger between the fingers or use a tweezer, you will squeeze the venom sack forcing more venom into the body.

stinger is torn from the bee's body. (See Figure 9-4.) For stings in the face area, you will need a mirror or the help of a friend.

Unfortunately, the author cannot recommend anything for the stings of bees, wasps, hornets, or yellow jackets that might be effective for everyone. The enzyme *papain*, found in papaya, seems to be effective for many. You might try making a salve of meat tenderizer that contains papaya and see if that works for you. Otherwise, try sympathy.

The bee that stings loses her life within a few hours, because the stinger is torn from her body, rupturing the intestine. Incidentally, the wasp, hornet, and yellow jacket can sting repeatedly, for their stingers are smooth; whereas the bee has a barbed stinger, which she leaves with the victim. If you can find a stinger after being stung, your attacker was a bee. If you cannot find a stinger, you were hit by something other than a bee.

UNIT 10

The Premedication Check

In the North: When the temperature for a part of each day reaches 60° F or more, the bees will become active and begin to fly out of the auxiliary exit and the hive entrance, if not blocked by dead bees and debris. This activity usually means your hive has made it through the winter. If you see no activity or very little activity after about ten days of pleasant warming weather (when the temperature is above 60°F for a part of each day), your hive may not have made it through the winter or may be very weak, if not dead. In this case, it is time to have a quick look inside the hive to determine its status—relatively strong, rather weak, very weak, or dead. You will want to quickly check the honey stores available. After replacing the hive top, remove the entrance reducer and scrape out the dead bees and debris so the bees can have free access to the hive entrance, then replace the entrance reducer.

In the South: Whenever the temperature is 60 to 65°F for a portion of the day during the winter, you can have a quick look inside if you suspect the hive is not prospering normally. Check the population, honey stores, and evidence of brood rearing, if any. You should make the following inspection, depending on your location in the southern area, between January 15 and February 15.

PERFORMING THE PREMEDICATION CHECK

Reread USDA Farmers' Bulletin #2255, "Identification and Control of Honey Bee Diseases." Know and look for the symptoms of American and European Foulbrood (AFB/EFB) each time you inspect the brood chambers.

Equipment needed:

☐ Hive tool
☐ Smoker or hive bomb

56 KEEPING BEES

- ☐ Frame grip
- ☐ Bee brush
- ☐ Empty super
- ☐ Long, thin stick

Step-by-step directions:

- ■ **Dress properly.**
- ■ **Light the smoker** or have the hive bomb handy.
- ■ **Enter the hive.**
- ☐ Remove frames 1 and 2 from the left side (or right side) of the upper brood chamber with the frame grip or hive tool. Place them in the empty super or stand in a safe place nearby.
- ☐ One by one, loosen, pick up, inspect, and move to the left (or right) side of the brood chamber frames 3, 4, and 5. Check the frames moved for bees, brood, honey stores, and symptoms of AFB/EFB.
- ☐ You will have seen one of the following during the previous steps:
  1. Numerous bees actively moving.
  2. A limited number of bees actively moving.
  3. A few or many bees just barely moving.
  4. No bees moving; perhaps just a moldy cluster of bees that drop off when touched or moved.

     By this time you will know that you have (1) a strong active hive; (2) a weak but active hive; (3) a hive that appears to be starving or suffering from disease; or (4) a hive that died out over the winter from winter kill, starvation, or disease. The hives that will need your immediate attention will be those in category 3, the bees just barely moving.

1. *If bees are numerous and active* and honey stores are available:
   - ☐ Be sure you have checked for symptoms of AFB/EFB.
   - ☐ Replace all frames in their original positions. You may have seen evidence of brood rearing beginning. Close the hive.
   - ☐ Remove the entrance reducer. With the long, thin stick pushed in the entrance, drag out as much of the dead bees and debris as possible by repeatedly dragging the stick across the bottom of the hive. You may want to set the brood chambers off of the bottom board onto a pair of two-by-fours or on the inverted hive top and really give the bottom a good cleaning with the hive tool. Replace the brood supers on the clean bottom. Replace the entrance reducer in the hive entrance.

2. *If bees are few and active* and honey stores are available:
   - ☐ Be sure you have checked for symptoms of AFB/EFB.

THE PREMEDICATION CHECK   57

- ☐ Replace all frames in their original positions. You may have seen some evidence of early brood rearing beginning. Close the hive.
- ☐ Remove the entrance reducer. With the long, thin stick pushed in the entrance, drag out as much of the dead bees and debris as possible by repeatedly dragging the stick across the bottom of the hive. You may want to set the brood chambers off of the bottom board onto a pair of two-by-fours or on the inverted hive top and really give the bottom a good cleaning with the hive tool. Replace the brood supers on the clean bottom. Replace the entrance reducer in the hive entrance.

3. *If the bees are few or many and barely moving* and with very little honey stores:

    - ☐ Replace all frames in their original positions and close the hive.
    - ☐ Immediately prepare a sugar solution as follows: Fill a quart jar ⅔ full of sugar (either cane or beet). Add hot water to the jar of sugar while stirring to dissolve the sugar. Continue to stir; when sugar is dissolved, fill the jar to the top with water.
    - ☐ Buy or clean up a trigger mist sprayer, like the ones many household cleaners are packaged and sold in. An alternate is a new paint brush that fits in the mouth of the jar of sugar solution.
    - ☐ Return to the hive. Smoke may not be necessary. Remove frames 1 and 2.
    - ☐ Gently remove the frames that have bees on them, one by one, and spray the bees with the sugar solution. If using a paint brush, dip it in the sugar solution and shake the solution on the bees.
    - ☐ As you spray the bees, check the combs for AFB/EFB. Replace the frames just sprayed in their original positions.
    - ☐ Remove a frame near the cluster of bees and either spray the solution into the comb or hold the frame at an angle and drizzle the solution onto the comb so the sugar solution will run into the cells. Repeat this step with a frame of comb on the opposite side of the cluster.
    - ☐ Return all the frames to their original positions. Close the hive.

    What you have done in the steps above is help the hive recover from the starvation mode by providing an immediately available source of food stores so the bees can fill their stomachs and begin to feed other weaker bees around them. Remember that a weak or starving hive cannot utilize granulated cane or beet sugar dumped into the hive or on the inner cover quickly enough to recover from starvation. Combs of honey that are not uncapped will not provide stores quickly enough to stop starvation. Starving bees will not have the strength to uncap honey stores given them or to salivate granulated sugar into usable nectar.

☐ Remove the entrance reducer. With the long stick, clean up the bottom board by repeatedly pushing it in the entrance of the hive and dragging out as much dead bees and debris as possible. You may want to set the brood chambers off of the bottom board onto a pair of two-by-fours or on the inverted hive top and really give the bottom a good cleaning with the hive tool. Replace the brood supers on the clean bottom. Replace the entrance reducer.

4. *If your bees are dead:*

☐ One by one, remove and brush the dead bees from each brood frame into something that can be carried away for burial if you find symptoms of AFB/EFB. Brush the bees from both brood chambers and scrape the dead bees and debris from the bottom board. Check the brood frames very carefully for symptoms of AFB/EFB. Call the bee inspector if you suspect AFB.

☐ *If the hive is free of disease:*

In the South: Store the brood chambers, frames and all, one each in a large plastic garbage bag with several tablespoons of Paradichlorobenzene (PDB). Tie the bag off securely and store it carefully until you can get a package or a swarm of bees to restart the hive.

In the North: Remove the entire hive into a shed that is bee-tight, or, lacking that, close up the hive entrance tightly so that other bees cannot rob out the hive before your package of bees arrives. If you are not going to restock the hive, it should be stored away with PDB moth crystals, as mentioned above.

Do not be concerned if the combs are moldy or if there are dead bees in many, many cells. When you restock the hive with a package or a swarm, those bees will clean up the comb until it shines like new.

☐ Order a new package of bees for delivery as soon as possible.
☐ Store the hive top, hive bottom, inner cover, and queen excluder (if you forgot to take it off last fall).

Dead or winter-killed hives should be checked very carefully for symptoms of foulbrood before being restocked with a package of bees. The supers should be aired for several days before the package is to arrive, if the brood supers have been stored in PDB crystals.

SOME TIPS ON CHECKING A DEAD OR DYING HIVE FOR FOULBROOD

Anytime you find covered brood in a dead hive, you must suspect AFB. The following tips may be of some help in sorting out situations in which AFB is probably not involved in the death of the hive.

1. If you find virtually no covered brood in the combs of the dead hive, it is not likely that the hive had AFB.
2. If you find dead bees that have crawled into the cells and very little capped or covered brood, it is a strong sign that the hive has starved to death. Many hives do starve to death after making it through the winter. As the bees start up brood rearing early in the spring (February or March in some of the most northern states), the hive begins to use up its limited stores quite rapidly, and without your help, the hive may starve to death just after getting started in the spring. In this case, you will find dead bees in the cells, probably some dead covered brood, and no honey stores.
3. The death of a hive during the spring, summer, or fall may have been the result of a swarm occurring or the beekeeper having killed the queen unknowingly during a manipulation or inspection involving the brood chambers. The bees will make a new queen, but the virgin queen may never return from her mating flight. Unless the beekeeper discovers the loss and re-queens in time, the hive will become a laying worker hive, and the next stage is complete destruction of the combs by the wax moth. If there is enough comb left when you discover the dead hive, it is hoped that you will find that all the brood has emerged except some covered brood with the characteristic protruding drone cell cappings. (See Figure 26-1.) In this case, the hive probably has not died from disease but from the loss of the queen and the inevitable loss of the hive without the intervention of the beekeeper.
4. Occasionally all the bees in a hive will abscond, leaving brood, honey stores, and pollen. In this case, it is probably not diseased, but since you will not know unless you see the bees abscond, you must assume that it is disease. You can always send a sample off for diagnosis to be absolutely sure any hive with a goodly amount of capped brood did not die of AFB. If you have reason to believe a dead hive may have disease, close the hive completely and do not let other bees have access to the hive until you confirm that it is healthy. If you suspect disease in a weak hive, keep an eye on it and close the entrance to one inch so it cannot be robbed out easily, until you confirm that the hive is healthy.

UNIT 11

Earliest Manipulations of the Hive

These manipulations and associated activities of the hive will begin as soon as the daily temperature reaches 60°F for a part of each day in the northern areas. For southern areas, where the bees are active most days during the winter, a few paragraphs will be added, where necessary, to indicate the approximate time to perform the earliest manipulations. Otherwise, the manipulations will be the same and will be performed with the same steps.

ABOUT TERRAMYCIN®, WHICH CONTROLS AFB/EFB

Terramycin® is most readily obtained as Terramycin Soluble Powder (TSP) Animal Formula in the 6.4-ounce packet. (See Figure 11-1.) It can be purchased from bee supply houses or shops and many feed stores where veterinary supplies are sold. A folder of instructions comes with the packet and now includes directions for medicating bees. Beekeeping authorities do not recommend medicating in a sugar solution but instead recommend using the dry mix and applying it to the frame tops of the brood chamber. Terramycin® loses its potency quickly in a watery solution, so if you use Terramycin® in a sugar solution, mix no more than the hive will consume in twenty-four hours. This handbook will not discuss Terramycin® in a solution but only as a dry medication.

The directions that come with (not *on*) the packet of Terramycin® will indicate the strength of the Terramycin® (TM) you have or are getting. Near the top of the first page of the directions you will see the statement, "Each pound contains XX grams of Oxytetracycline Hydrochloride." The number in that statement tells you whether it is TM-5, TM-10, TM-25, TM-50, or the more recently available TM-100. You have a better-than-average chance of obtaining TM-25 when and wherever you purchase Terramycin®. This number means you will use five times as

Figure 11-1, Medicines for Bee Diseases. The 6.4-ounce packet of Terramycin® and the 0.5-gram bottle of Fumadil-B® are the smallest quantities usually available. Terramycin® controls AFB/EFB, and Fumadil-B® controls Nosema Disease.

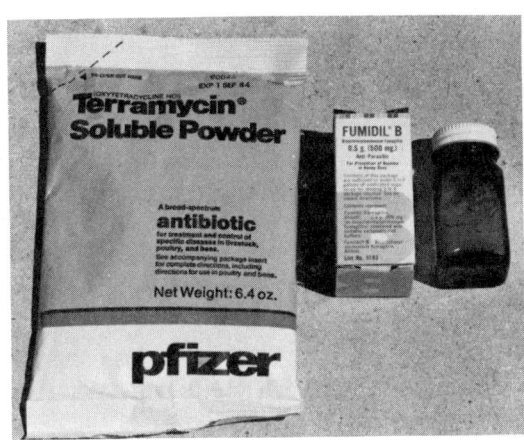

much TM-5 in preparing your medication as you will TM-25, or half as much TM-50 as TM-25. The directions given in this book are for TM-25, because this is the most common strength of Terramycin® sold. Each teaspoonful of TM-25 contains 200 milligrams of Oxytetracycline, which is the recommended amount to be mixed with at least one ounce of powdered or Drivert® sugar for each medication of the hive.

Procedures given in this volume for the preparation and dispensation of the medication will vary from the directions that come with the packet of Terramycin® as follows: (1) Medication is prepared for the full year for one (or more) hive; (2) more sugar is used in the medication; and (3) the medication is dispensed at seven-to-ten day intervals.

The logic of the medication plan is this: (1) You mix your medicine once each year; (2) you use three ounces of sugar for each 200 milligrams of Oxytetracycline to disperse the medication over a larger brood area for feeding the larvae; and (3) the medicine is dispensed at seven-to-ten day intervals to keep the medication over a larger brood area for a longer time as the brood rearing increases in the spring. This medication is also spread over a larger area of the brood frames if the hive is being fed or medicated for Nosema Disease at the same time that the Terramycin® mix is being applied on the frame top bars.

Preparing the AFB/EFB Medication

This medication is for the control of AFB/EFB diseases. The steps below outline the preparation of medication for one hive for one year. Multiply the ingredients by the number of hives you will be medicating, and increase the size of the jar for mixing and holding the medication.

- [ ] Place one pound of powdered sugar or Drivert® sugar in a larger-than-adequate-sized jar.
- [ ] Place five level teaspoons of TM-25 in the jar with the sugar (25 tsp of TM-5, 13 tsp of TM-10, 2½ tsp of TM-50, or 1¼ tsp of TM-100).
- [ ] Screw the cover on the jar, and mix thoroughly. Label the jar and store out of the reach of children.

The above mixture for one hive for one year consists of about 35 level tablespoons (more if you use TM-5 or TM-10) to be dispensed over five medications during the year, or seven level tablespoons per medication.

Before medicating the hive, prepare the pollen supplement if you will be placing a pollen supplement patty on the frame tops and spooning the Terramycin mix around it.

POLLEN SUPPLEMENT

Pollen supplement feeding is essential in some areas for early brood rearing if natural pollen is not available in the field in adequate quantities. The same can be said for nectar. You can provide for both needs by placing pollen supplement patties on the frame tops and feeding sugar solution. Pollen supplement usually contains a small percentage of natural pollen, while pollen substitute contains no natural pollen. The pollen supplement powder is a mixture of brewer's yeast, powdered milk, and soy flour. Pollen supplement is only a substitute for natural pollen to span a period of a few weeks until natural pollen is available in the field. When natural pollen is available in adequate quantities, the bees will not use the pollen supplement.

If you add the pollen supplement patty at the time of your first AFB medication, you can see how much has been used in seven to ten days at the time of the second AFB medication. If very little has been used, medicate around it and and check it on the third medication. If the bees have not used the pollen patty by this time, it will be hard and moldy; use the hive tool to toss it out. If the pollen patty has been used up (or nearly so), medicate, and make more pollen supplement patties available on the frame tops as soon as possible. If the bees reject the pollen patties for a couple of years, you can assume there must be enough early natural pollen available to take care of the hive needs, and you can stop feeding pollen supplement.

Pollen traps are available that permit you to trap away pollen during the honey flow till fall (one day out of four) to feed back to the bees (as pollen patties) in the early spring rather than using pollen supplement.

EARLIEST MANIPULATIONS OF THE HIVE

Bees do not reject natural pollen but will use it even though there is adequate pollen in the field. (See Unit 34.)

Do not purchase the pollen supplement with Terramycin premixed in it *until* you determine that your bees will take the pollen supplement and in so doing be medicated. If your bees do not use the medicated pollen supplement, they will not be getting the medication.

Preparing the Pollen Supplement Patty

- ☐ The night before you plan to medicate for AFB/EFB, mix the pollen patty as follows:
  Note: These ingredients will make one patty for one hive. Multiply ingredients by the number of hives to be fed.
- ☐ Mix ¼ cup sugar with ⅛ cup hot water. Dissolve the sugar completely.
- ☐ Place one cup of pollen supplement in a mixing bowl.
- ☐ Add half of the sugar solution to the pollen supplement and stir.
- ☐ Continue to add sugar solution until the pollen supplement mixture has a consistency similar to peanut butter cookie dough. Add more pollen supplement if the mixture is too thin.
- ☐ Dump the mixture out on a sheet of waxed paper, and press it into a patty about ⅓ inch thick. Press a sheet of waxed paper on top of the patty.
- ☐ Let the patty set in a safe place, on a flat surface, until you are ready to place it on the frame tops the next day.
- ☐ Label and store the remaining pollen supplement powder in a tightly sealed jar to keep out vermin. If the bees use up the pollen patty quickly, mix another batch or make a double batch.

MEDICATING TO CONTROL AFB/EFB AND FEEDING THE POLLEN PATTY

Equipment needed:

- ☐ Hive tool
- ☐ Smoker or hive bomb
- ☐ Pollen supplement patty
- ☐ Terramycin® mix

Step-by-step directions:

- ■ **Dress properly.**
- ■ **Light the smoker** or have hive bomb handy.
- ■ **Enter the hive.** Smoke the frame tops of the upper brood chamber

to move the bees down so they won't be crushed when you place the pollen patty on the frame tops.

☐ Remove the waxed paper from the top of the pollen patty and flop it over onto the frame tops of the upper brood chamber. The waxed paper, now facing up, will help to keep the patty from drying out too quickly. (See Figure 11–2.)

☐ Spoon seven level tablespoons of the Terramycin® mix on the frame top bars around the pollen patty. It doesn't matter if some falls between the frames.

☐ Replace the inner cover *with the rim side down* to make room for the pollen patty. Close the hive, and begin medicating to control Nosema Disease. (See "Medicating to Control Nosema Disease" in this unit.) Label and store the remaining Terramycin® mix in a cool, dry place out of the reach of children.

☐ Wait seven to ten days and repeat the steps above. Do not make another pollen patty until you check the patty in the hive. If it has been used, then make a patty later and add it. In the meantime, spoon the Terramycin® mix on the frame tops and close the hive. If the pollen patty has not been used, spoon the Terramycin® mix on the frame tops around the patty and close the hive.

☐ Wait seven to ten days and repeat the steps above. If the bees have not used the first patty, toss it off the frame tops with the hive tool. If the bees are still taking the pollen supplement, add another patty. Continue to add pollen patties until the bees will not use any more.

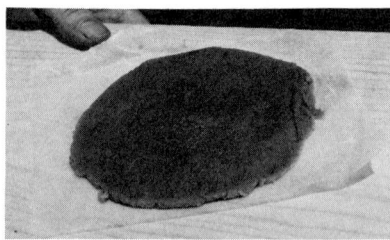

Figure 11-2, Dispensing AFB/EFB Medication around a Pollen Patty. Above: A pollen patty with the waxed paper removed from one side, ready to be flopped over onto the frame tops. Right: Medicating around the pollen patty. The waxed paper is left on the pollen patty to keep it moist longer. The bees will tear away the waxed paper as they use the pollen patty.

PREPARING THE SUGAR SOLUTIONS FOR FEEDING AND MEDICATING

Never buy honey to feed bees. Do not feed your honey to other bees in your apiary unless you medicate all hives with a regular medicating routine, as will be discussed later. AFB/EFB and Nosema Disease are spread in honey from diseased hives. It is best to use either cane or beet sugar for feeding and medicating your hives. For medicating to prevent and control Nosema Disease, the directions call for a two-part sugar to one-part water solution (2-1 solution) whether fed in the spring or fall. For additional feeding in the spring after the Nosema medication, you will use a 1-1 solution.

Steps for preparing one gallon of 2-1 solution:

- ☐ Mark the one-gallon capacity in a six-quart or larger pan.
- ☐ Place five pounds of sugar in that pan.
- ☐ In a separate pot, bring two quarts of water to a boil.
- ☐ Pour at least one quart of the boiling water on the sugar, and stir to dissolve. Continue to add hot water and stir until you reach the one-gallon mark. Stir in $\frac{1}{16}$ teaspoon cream of tartar to prevent crystallization of the solution.
- ☐ Let the solution cool. Pour it in a jug or container, label it, and store it until needed.

    This solution will be used for spring and fall Nosema medication or for fall feeding, if necessary, to bring a hive up to wintering condition by providing quick, high-sucrose food stores.

Steps for preparing one gallon of 1-1 solution:

- ☐ Mark the one-gallon capacity in a six-quart or larger pot.
- ☐ Place ten cups of sugar in that pot.
- ☐ In a separate pot, bring three quarts of water to a boil.
- ☐ Pour at least two quarts of the boiling water on the sugar, and stir to dissolve. Continue to add water and stir until you reach the one-gallon mark.
- ☐ Let the solution cool. Pour it in a jug or container, label it, and store it until needed.

Steps for preparing the Nosema Disease medication:

- ☐ Prepare one gallon of the 2-1 solution, let it cool, and pour it in a jug.
- ☐ Shake the vial of Fumadil-B® thoroughly. The idea is to unpack the powder in the vial. (See Figure 11-1.)
- ☐ Place one level teaspoon of Fumadil-B powder in the gallon of 2-1

solution. Cap the jug and shake to mix the powder in the sugar solution. Mix thoroughly.

Note: It may be best to shake the teaspoon of Fumadil-B powder in a small jar with a small quantity of the solution in order to get it mixed thoroughly. Then pour it into the jug and shake again to mix the medicine thoroughly.

☐ Label and store the Nosema-medicated solution out of the reach of children.

MEDICATING TO CONTROL NOSEMA DISEASE

Equipment needed:

☐ Hive tool
☐ Smoker or hive bomb
☐ Entrance or Boardman feeder
☐ Quart jar
☐ Gallon jug of Nosema-medicated sugar solution (2-1 solution)

Step-by-step directions:

■ **Dress properly.**
■ **Light the smoker** or have hive bomb handy.
☐ Smoke the hive entrance, and insert the entrance feeder base. Fill the quart jar with the Nosema medication from the gallon jug. Tighten the feeder lid securely, and invert the quart jar over and into the feeder base.
☐ Each time the jar empties to about one inch from the bottom, remove and refill. Between uses, store the jug of medication out of the reach of children. *Caution: Avoid cuts* by using a rag or glove when unscrewing the feeder cap from the feeder jar.
☐ Repeat the previous step until the gallon of Nosema-medicated sugar solution has been used up.
☐ If additional feeding beyond the Nosema medication is required, prepare and continue to feed the 1-1 sugar solution.

SETTING UP A CONSTANTLY AVAILABLE WATER SOURCE

Many backyard beekeepers in urban areas make the mistake of not having a constantly available water source for their bees. The bees may become a nuisance by going to the neighbor's bird bath, fish pond, or swimming pool. Bees tend to train to a water source if started early in the spring and if it should dry up will retrain to a new source. Once the bees have trained to your neighbor's yard, it will be difficult to retrain them back to your yard. You must remain vigilant so that your water source does not dry up. Bees prefer a watering source such as a creek, stream, irrigation canal, pond, drainage ditch, spring, or seepage spring rather than water in a pan or tub, if the natural system is within one-half mile. The following man-made systems work well if you keep the water source constantly available.

1. In a pail, tub, bird bath, etc., float two-by-fours or add rocks, then drape an old towel over them and in the water. The towel in the water acts as a wick and remains damp. The bees land on the damp towel to drink without fear of drowning. (See Figure 7-1.)
2. A piece of rough-cut lumber can be propped up at one end and a water hose laid at the highest point on the inclined board with the hose just barely dripping. As the water trickles down the board, the bees can land and sip water from the damp fibers of the rough-cut lumber. This works well and can be placed so that the constant trickle of water down the board can be used to water a tree or plant over an extended period of time.

Bees require water in quantity in the early spring to water down the honey stores in the brood chambers to a nectar consistency for feeding the larval brood. As the nectar flow becomes heavier, the nectar is fed directly to the larvae, and water is not required in such large quantities.

Water is required in quantity during the hottest part of the summer for evaporative cooling to keep the hive, both brood and honey supers, at a temperature of about 94°F. If water is not available to the bees for this cooling, the combs may soften and break down under the weight of the honey and brood, or the brood may become overheated and die.

Evaporative cooling in the hive works as follows: When more cooling is needed beyond the drawing in of outside air and the hanging out of bees, the field bees stop gathering nectar and begin gathering only water. It takes a bee about forty to seventy seconds to tank up at the water source plus the time it takes to fly to and from the source. Inside the hive, the water is strategically placed on the frame tops and other areas. As the bees in the hive draw in the hot, dryer air, the water evaporates and cools the hive. (As water evaporates, it absorbs heat and cools the area.)

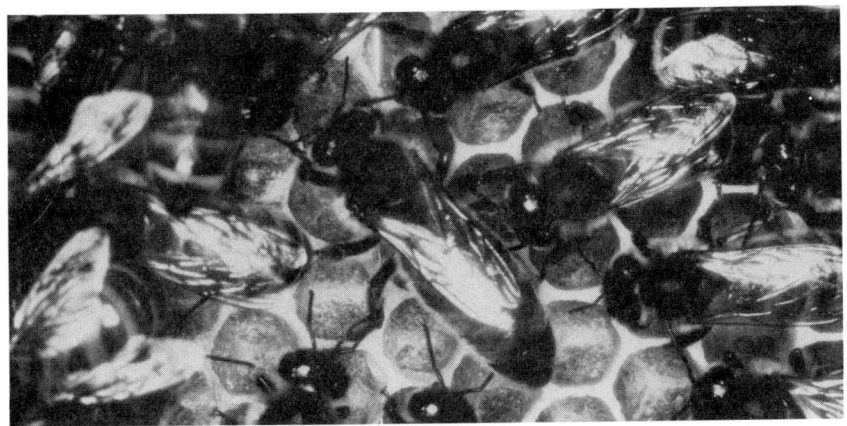

PART **IV**

SPRING ACTIVITIES AND MANIPULATIONS

The next six units cover the activities and manipulations that will or may be necessary to get the new hives started, strengthen weak hives, and stop strong hives from swarming. In general, these activities will be performed after spring weather has become somewhat settled, beginning about the time of early fruit bloom in your area. Planning will have been done earlier if you are re-queening or starting new hives with package bees or swarms.

- **Unit 12** INSPECTING THE BROOD CHAMBER
- **Unit 13** RE-QUEENING THE HIVE
- **Unit 14** INSTALLING A PACKAGE OF BEES
- **Unit 15** SWARMS AND SWARM CONTROL TECHNIQUES
- **Unit 16** CAPTURING SWARMS
- **Unit 17** WHAT TO DO WITH SWARMS

Unit **12**

Inspecting the Brood Chamber

As you gain more experience working with bees, you may not inspect the brood chambers of established hives as often as the newly established hives or hives that you know have swarmed. You should make a minimum of three brood chamber inspections of all established hives during the beekeeping year.

The three inspections should be these:

1. In early spring (temperature range 65 to 70°F), verify that you have a queen. The brood pattern will indicate whether the queen should be replaced by a new mated queen. This inspection could be done at the time of your final medication in the spring.
2. Inspect at the time you add your second honey super. Remove the first honey super and queen excluder, and check on the queen and the brood pattern. This check, usually after the swarm season, will alert you to the fact that your hive may have swarmed, the new queen may not have returned from the mating flight, and you must re-queen. It is nice to know that you have a queen going into the rest of the honey flow, because you will not want to remove heavy honey supers to check on the queen unless it appears that something is wrong. Do not hesitate to inspect the brood chamber at any time if you suspect something is wrong. If you suspect, then confirm.
3. Inspect early in the fall, as soon as you have removed all honey supers for extracting or storage. Check on the queen and the brood pattern to determine whether to re-queen as soon as possible or go through the winter with your present queen. It is always good to know that you are going into the winter with a queen. This inspection also gives you the opportunity to check honey stores in the brood area.

You should always inspect the brood chamber three weeks after you know your hive has swarmed. Do not expect every hive to re-queen automatically. About 25 percent of the time, a hive that swarms does not re-queen. If for some reason the virgin queen does not return from her mating flight, the hive is left hopelessly queenless unless you discover there is no queen and re-queen.

Check your brood chamber if you suspect the hive is not prospering in population growth or in honey production during the normal honey flow period in your area. If you have several hives and one is not keeping up with the others, inspect that hive.

If on inspection of a slow hive you find eggs and larvae, suspect disease or skunks. Skunks enjoy eating bees. A number of skunks around can keep a hive population low enough to be not at all productive but just maintaining itself.

Of course, you can inspect the brood chamber as often as you wish. Most experienced beekeepers feel the less they interfere in the brood area, the better the hive will thrive; but *do* make the three suggested inspections each year. Underinspecting brood chambers is worse than overinspecting them.

To inspect the brood chamber, enter the hive and remove one or two side frames. Loosen the ends of the remaining frames so you can move them aside until you reach the frame in which you are interested. Removing the side frame or frames and storing them in an empty super or other safe place leaves a space into which you can move the next frame. You can then lift it up and out with little chance of damaging the queen. As each frame is replaced and pushed to the side against the other frames, you have space in which to remove the next frame. You can move several frames aside, one after the other, until you reach the frame of interest to you. Center it in the space, and remove it for inspection.

In wild hives, bees use brace comb to fasten free-hanging combs in the hollow tree to the side of the cavity and to each other so that they do not swing as the tree sways in the wind. The bees also use brace comb in the hive, and it can be a nuisance when you want to inspect the frames in a movable-frame hive. For example, the bees often add brace comb from the side of the super to the side of the nearest comb. The bees also place comb (called burr comb) between the frame top bars of most of the frames, connecting them together. Some hives may add so much comb between the frame top bars that it nearly closes them, leaving access to the supers above only through several small openings between each pair of frames. This is a nuisance, but if you know it may happen, you can plan ahead for it when you go into the brood or honey supers. Proceed as follows if you find brace comb connecting adjacent frame top bars.

INSPECTING THE BROOD CHAMBER   73

- [ ] With the hive tool, slice through the comb along the side of each frame top bar *opposite* the one you want to remove. This leaves the brace comb attached to the frame you will remove. (Figure 12-1.)

- [ ] Loosen the frame to be removed on each end with the hive tool. Use the frame grip to lift the frame up and out of the super. The brace (or burr) comb comes out on the frame top of the frame you lift out. If you cut the burr comb along the top of the frame you want to remove, the burr comb will remain on the frame tops in the hive, gouging and ripping the cappings or the comb of the frame you lift out. It could possibly roll and kill bees, perhaps the queen. It might also start robbing in the apiary, for those gouged combs will leak honey as they stand nearby or hang in the empty super during your inspection of the brood or honey supers. The spilled honey around the hive could start robbing after you close up the hive.

- [ ] To begin inspecting the brood chamber of established hives, first remove the second frame from one side of the brood chamber. The reason is this: If the bees have placed brace comb from the side of the super to the side of the nearest comb, the brace comb will gouge and rip the cappings and comb as you lift it out of the super. By starting with the second frame from the side, there is much less chance of this happening.

- [ ] Once the brood frame you have chosen to inspect is out of the brood chamber, grasp each end of the frame top bar with the thumb and forefinger of each hand. Lift the frame horizontally to eye level, and turn so the light comes from behind you. Push the buttom of the frame toward you with your middle or fourth finger against the end bars until you can see directly into the bottom of the cells. (See Figure 12-2.)

- [ ] On these inspections you will not be particularly interested in searching for or finding the queen, but in finding eggs in the brood comb cells and observing the laying pattern of the queen. The laying pattern should show that she is laying in almost every cell available. Finding an egg in each cell *(This is important—one egg in each cell!!)* tells you that your queen is there and laying. Because the eggs hatch in three days, you know that your queen was in the hive within the last three days if you see eggs. Finding two or more eggs in any cell indicates you have no queen in the hive but the hive has laying workers.

- [ ] To check the laying pattern, find a frame containing mostly covered brood or a large area of covered brood. A quick look will tell you if most cells are capped, and this is what you want to know. If there

Figure 12-1, Removing Frames Connected with Burr Comb between the Frame Top Bars. Top: The frame to be removed first is the one between the two hive tools. Notice that the burr comb has been cut through on the sides of the frames adjoining the frame to be removed. Bottom: You can see that the burr comb is being brought out on the top bar of the frame being removed.

Figure 12-2, Inspecting Frames of Brood. Turn to have the light coming from behind you as you tilt the frame slightly so you can see the eggs and tiny larvae in the bottom of the cells. In these photos, notice how the frame grip hangs on the frame when a rubber band is used to keep the frame grip closed over the frame top.

are more than about forty or fifty uncapped cells in amongst the capped cells, this is a good indication that your queen is old or failing. You may want to replace her before fall or as soon as possible.

The first early inspection may not be the best one on which to evaluate the laying pattern of the queen, for this reason: In the cooler weather of fall with less nectar coming in, the bees contract the brood area on the brood frames by filling the cells with both nectar and pollen. It is possible to have a good queen who begins laying but finds cells still containing pollen or honey. The queen will bypass these and move on to other cells and other frames. If you inspect three weeks later (one full metamorphic cycle), you will have a better feeling about whether your queen is a good layer, filling almost all available cells as she goes along.

☐ Check both sides of the brood frame when you are inspecting. Here is how to get the back side of the frame into view without becoming a contortionist: You are presently holding the frame top bar horizontally by each end. When you are ready to look at the back side, drop one arm until the frame top is vertical. Rotate the frame 180° (by manipulating the fingers), then raise your arm up to a horizontal position. Your frame is now upside down and facing you for inspection. In this position the frame must be tipped away from you, supported by a finger behind each end bar, so you can look directly into the bottom of the brood cells. Practice this a few times with an empty frame until it becomes easy for you to do.

☐ This first inspection is primarily to determine whether you have a queen and that she is laying or whether you have no queen and perhaps have laying workers. If you have a queen, everything is fine for now. If you plan to re-queen this spring, you must de-queen (remove the existing queen) before introducing your new mated queen. If you have no queen, you should order a new queen immediately and introduce her as soon as she arrives to get the hive moving ahead with brood rearing. If you find multiple eggs in many cells, you have laying workers and more of a problem. Refer to Unit 26, which covers laying workers.

How can you tell if you have no queen? (1) If you find no eggs, no larvae, and no covered brood, you probably have no queen and should order a queen immediately. (2) If you find no eggs, no larvae, and only covered brood, you probably have no queen. You can either order a queen immediately or wait one week and inspect the brood chamber again. If you still find no eggs, no larvae, and less covered brood, then order a queen immediately.

Beware: If you spot eggs, *always* check carefully to make sure there is only one egg in each cell, because multiple eggs in some or many cells indicates laying workers. Laying worker hives are very hard to re-queen, for the bees think they have a queen and will not accept a new queen when one is introduced in the usual manner. You will learn from experience that the queen lays in almost every cell as she moves along, whereas laying workers lay in no pattern but lay the multiple eggs in a helter-skelter manner. (See Figure 26-1).

If you plan to add a frame of brood containing eggs and larvae to let the queenless hive make a new queen, *do not* begin too early in the season, because there may not be drones available for mating with the virgin queen. Wait until you see a fair number of drones around before you try this re-queening method. *Remember* to check the brood chamber three or four weeks after introducing the frame with eggs and larvae to make sure the new virgin queen has mated and begun to lay. Buying a mated queen, if available, usually gets new brood rearing started two to three weeks earlier than rearing a queen from eggs and larvae in the hive and avoids the risk of the virgin queen not returning from the mating flight. The purchased queen is mated and ready to begin laying.

■ INSPECT THE BROOD CHAMBER

Equipment needed:

- [ ] Hive tool
- [ ] Frame grip
- [ ] Smoker or hive bomb
- [ ] Empty super

Step-by-step directions:

- ■ **Dress properly.**
- ■ **Light the smoker** or have hive bomb handy.
- ■ **Enter the hive.** You may not want to smoke the entrance. Invert the hive top on the ground nearby, and set the empty super across it.
- [ ] Remove one or two frames from one side of the brood chamber with the frame grip or hive tool. Place the frames in the empty super.
- [ ] With the hive tool, loosen frames 3 and 4. Push both to one wall of the brood chamber.
- [ ] With the hive tool, loosen frame 5 and slide it to the center of the open space in the brood chamber. With the frame grip, remove frame 5 by lifting it straight up and out of the brood chamber. Grasp one end of the frame top bar with your thumb and forefinger, release the frame grip, and grasp the other end of the frame top bar. If you have no frame grip, grasp each end of the frame top bar with the

- thumb and forefinger of each hand, and lift the frame straight up and out of the brood chamber.
- ☐ With your thumbs and forefingers grasping each end of the frame top bar, turn so the light comes from behind you, and lift the frame to eye level. With a finger behind each end bar, push the bottom of the frame forward (toward your body) until you are able to look directly into the bottom of the cells.
- ☐ First, look for the larvae that are easy to see as white grubs in the cells. When you spot the larvae, look outward in any direction toward the edges of the frame. You should see smaller and smaller larvae, then the egg in the cells. Make sure there is only one egg in each cell. (See Figure 12-3.)
- ☐ Turn the frame over and look at the other side. Holding the frame top horizontally, lower one hand until the frame is vertical, rotate the frame top 180° (by manipulating the fingers), then raise the frame top bar to the horizontal position by lifting the lowered hand. The frame is now inverted; tip the frame away from your body slightly, supported by a finger behind each frame end bar, until you can look into the bottom of the cells.
- ☐ When you find eggs, you will know that your queen is there. Now evaluate the laying pattern of the queen. Check for symptons of American Foulbrood Disease. (See Unit 8.)
- ☐ Unless the frame you are holding has a fairly large area of covered brood, replace it and push it against the other frames on the side of the brood super. In turn, loosen, center, lift out, and look at frames 6 and 7. Look at the covered brood on the frames, and determine if there are more than about fifty empty cells scattered among the covered brood. (Give extra credit if there is a larva in each of the uncapped cells.) If there are a good many empty cells, you may want to consider re-queening.
- ☐ As you look at the covered brood, check for the symptoms of AFB/EFB. (see Unit 8 or reread Farmers' Bulletin #2255.) If you wish to check the color of the capped brood, lift several cappings gently with the edge of the hive tool point or a pen knife. Do not worry about uncapping a few pupae. The uncapped pupae will not die (if they are healthy); in fact, occasionally you will find a hive that may not cover many of the pupae (called bare or bald-headed brood). This is probably a hereditary trait. If you find this occurring in your hives, consider re-queening to eliminate the lazy gene from the species.
- ☐ If you may be spending some time inspecting the hive, set the inner cover over the brood chamber to keep the brood from cooling too much. To get another frame of brood, stand the inner cover against

INSPECTING THE BROOD CHAMBER 79

Figure 12-3, Eggs and Larvae. These photos show in close-up the eggs and larvae in the cells of the brood frame. Spotting the eggs as seen here (one egg per cell) tells you that your queen is in residence and is working.

your leg (bee side away from the leg, of course), replace the frame you are holding, pick up another frame, and replace the inner cover.
☐ When you have completed your inspection, replace all frames in their original positions. Close the hive by replacing the inner cover and hive top, or proceed with whatever manipulation is required before closing the hive.

UNIT **13**

Re-Queening the Hive

Before discussing the steps for the manipulations of re-queening, consider the occurrences that make re-queening necessary. You should be aware that re-queening as used at this point in the discussion may entail two manipulations, one following the other. These are de-queening and then re-queening.

Finding and removing the old queen in a hive is called *de-queening*. De-queening is necessary if a queen is in the hive at the time you want to re-queen. The bees will not accept a new queen if a queen is already in the hive or if queen cells are being developed in the hive. De-queening usually must be done before spring or fall re-queening or to re-queen a mean or cross hive.

Re-queen after the old queen has been removed or if no queen is in the hive. (The lack of a queen can be caused by the loss of a virgin queen after the hive has swarmed or the loss of a queen during a manipulation of the hive.)

Do not de-queen the hive until you have your new mated queen in your hands. If you de-queen one or more days before your new mated queen is introduced into the hive, the bees will have quickly realized that they had no queen and will have begun to develop queen cells. Your chances of the bees accepting the new queen will be drastically reduced unless you inspect and tear out *all* queen cells before introducing the new mated queen.

Always inspect the brood chamber before introducing a new mated queen. If you discover two or more eggs in any cell, it is almost certain that you have laying workers. A laying worker hive is almost impossible to re-queen in the normal manner described in this section. (See Unit 26.)

Always check your hive at least three weeks after the hive has swarmed to determine that the virgin queen has mated and begun to lay. If the inspection shows no eggs or small larvae, you can suspect that the queen has been lost on the mating flight and be prepared to order a new mated queen for the hive. (See Unit 4.) A few words of caution may save the beginner the cost of a queen that may not be needed. On many occasions a new queen raised in the hive may not have begun to lay enough eggs for the beginner to see or find after three weeks. The beginner, thinking the hive is queenless, will order a new mated queen and find several frames of eggs and small larvae when introducing the new queen. If you find no eggs or small larvae after three weeks, wait one more week before ordering your queen. If no eggs or larvae are found at that time, you can be fairly certain the hive will be queenless.

As suggested earlier, the brood supers should be inspected when you add the second honey super. The reasons for inspecting the brood chamber at that time are to determine that the hive is queenright going into the honey flow, and to make sure that if the hive swarmed without your knowledge it has re-queened itself and has a laying queen. If you discover the queen is missing at this inspection, order a new mated queen as soon as possible. Be sure the hive has not become a laying worker hive before ordering the new mated queen. (If the hive has laying workers, see Unit 26.)

Let us now assume you have the new mated queen in your hand and have found that she is alive and well, and you are ready to begin the manipulations to re-queen the hive. Do not leave the queen sitting out in the sun while you perform the preparatory manipulations, because she may not be alive and well when you are ready to introduce her into the hive. (See Figure 4-1.)

Some beekeepers think that the attendant bees accompanying the queen should be released from the cage before introducing the queen to the hive. They believe the animosity of the bees in the hive toward the strange bees may cause the queen to be stung in the cage or at the time of her release. If you think that this could happen, take the queen cage into the bathroom, turn on the light over the medicine cabinet, and open the end of the cage without the candy plug. Let the worker bees out one at a time while holding a finger over the end of the cage. In the event the queen should escape during this exercise she cannot fly away, because she is confined to the room and should be attracted to the light reflected in the mirror. You can hope to capture her without damaging her. The author feels that the risk of damaging the queen in releasing the attendants is often greater than the risk of introducing the queen cage with all attendants still in the cage. Be very gentle with the queen, and do not pick her up by squeezing the abdomen. You may injure her and cause her to die. In this, as in most things, experience in releasing the bees from the queen cage without releasing or injuring the queen comes with practice. When the queen is back in the cage, replace the cork.

With the cardboard queen cage, take the queen cage into the bathroom, turn on the light over the medicine cabinet, and release the bees on the mirror by sliding the outer cover from the inner box and lifting away the inner box. When the queen is alone, place the inner box over her. Gently slip the outer cover over the inner box, and you will have your queen alone in the cardboard cage.

DE-QUEENING THE HIVE

Two methods for de-queening the hive are described in this section. (1) De-queening by searching the frames for the queen, and (2) dumping the bees from the brood chambers to find the queen.

If you are certain that the hive does not have a queen in residence, skip to the section entitled "Re-Queen the Hive" in this unit.

Searching the Frames for the Queen

Equipment needed:

- ☐ Hive tool
- ☐ Smoker or hive bomb
- ☐ Frame grip
- ☐ Bee brush
- ☐ Empty super
- ☐ Pair of 18" two-by-fours
- ☐ Small jar with lid or a queen catcher (See Figure 16-5.)

Step-by-step directions:

- ■ **Dress properly.**
- ■ **Light the smoker** or have hive bomb handy.
- ■ **Enter the hive.** Invert the hive top on the ground nearby. Place the empty super on the inverted hive top.
- ☐ At any step in these instructions when you find the queen, herd her into the jar and close it, or catch her in the queen catcher. Replace any frames, replace the brood chambers, and skip to the section entitled "Re-Queen the Hive" in this unit.
- ☐ Set a pair of two-by-fours about twelve inches apart on the ground nearby.
- ☐ With the hive tool, lift and smoke between the honey supers (if any) and the brood supers. Wait one minute. Remove the honey supers one by one, and stack them across the two-by-fours.

84   KEEPING BEES

☐ If there are no honey supers, set the upper brood chamber across the pair of two-by-fours. Skip the next step.

☐ If there are honey supers stacked on the two-by-fours, gently remove the queen excluder and set it on top of the honey supers. Set the upper brood chamber on the queen excluder.

■ **Inspect the upper brood chamber.** (See Unit 12.) If there are eggs and larvae and there is only one egg in each cell, and if there are no multiple eggs in any cells, continue with the following steps to de-queen the hive. If there are multiple eggs in any cells, see Unit 26 on laying workers.

☐ Check frames 1 and 2 carefully for the queen. Place them in the empty super. Remove frames 9 and 10. Check them carefully for the queen. Place them in the empty super with frames 1 and 2. The queen is not usually found on frames of honey, and in most cases, frames 1, 2, 9, and 10 will contain honey.

☐ Check for the queen on the side walls of the super. Loosen and move frames 3 and 4, centering them one against the other in the space on that side of the brood chamber.

☐ Loosen and move frames 7 and 8, centering them one against the other on that side of the brood chamber.

☐ Frames 5 and 6 will be together in the center of the brood chamber. So far you have three pairs of brood frames spaced in the brood chamber so that the queen cannot move from one pair to another or to the side walls of the brood super. The queen should be on one of the three pairs of brood frames, probably between one of the pairs, because queens prefer the dark.

☐ Start with any pair. As you lift out one of the frames of the pair, look at the inside of the frame remaining and you may see the queen on that frame. If not, carefully inspect the frame you have removed, replace it, and remove the second frame of the pair to inspect it carefully for the queen. If you wish to move the bees from an area at which you want to look, blow gently on the bees through your veil. Experience is the best teacher for learning to spot the queen, but the following tips may help.

- Look for the characteristic ring of bees facing the queen.
- Look at any small, open areas on the comb not covered by bees. The bees usually make extra room for the queen.
- Check the edges and corners where the comb may have been cut away.
- Gently blow on thick masses of bees to get them to move, perhaps uncovering the queen.

☐ Repeat the previous step with each pair of frames. Keep the pairs together as you put them back. If you do not find the queen on this first inspection, go to the lower brood chamber and perform all the previous steps. If you do not find the queen, inspect all the frames of both brood supers again. If you do not find the queen on this second inspection, replace all the frames, close the hive, and try to find the queen later in the day or the next day. You might also try the procedure described next.

Dumping the Bees to Find the Queen

First you will learn how to shake the bees from the frames when the time comes to dump the bees. Brushing should not be necessary, because you are interested in dislodging the queen, who is not able to hold on as tightly as the workers.

Shaking bees from the frames may be necessary in other manipulations. Follow these same instructions when directed. You may be requested to shake the bees from the frames onto the top of the brood chambers or in front of the hive.

Shaking the Bees from the Frames

- [ ] When the frame to be shaken has been loosened and removed from the super, grasp each end of the frame top bar with your thumb on top of the frame top bar on each end and three fingers under the end of the frame top bar on each end. Move to the area where the bees are to be shaken.
- [ ] Push your hands downward rapidly about one foot, and stop the descent abruptly (be sure to have a good grip on the ends of the frame top bar). The bees will not expect the quick stop, and most will fall from the frame with the queen, if she is there. Place the frame that was shaken in an empty super, or stand it on end in a safe place nearby.

Why Dumping Works: When an actively laying queen, fat with eggs, is dropped or somehow gets outside of the hive, many of the bees from the hive soon find her and form a ball around her, presumably for protection. An actively laying queen cannot fly for any distance and if dropped will flutter to the ground nearby. This activity of forming a ball around a queen who cannot fly seems to be one of the behavior traits of the bees; you are taking advantage of this trait when you dump the bees (and queen) to find the queen for de-queening.

Dumping the bees is the easiest way to find the queen. In southern areas, where queens are not readily available for re-queening until the hive is teeming with bees and has produced many drones for the expected swarm season, looking for the queen on the very crowded frames can become a chore and may take several attempts. Some beekeepers suggest placing a piece of excluder over the hive entrance so the queen cannot get back into the hive. A better method is to use no entrance excluder but to dump the bees farther from the entrance so the queen has farther to run to get back to the hive. It helps if you have another pair of eyes watching while you perform the dumping manipulation. One reason for not using the entrance excluder is that it slows down the traffic going back into the hive. The existence of many drones virtually stops all traffic into the hive.

Note: If you ever find a ball of bees about the size of a walnut or golf ball outside around the hive, you can be pretty sure the queen has been dropped or has somehow gotten outside the hive. If your queen has a clipped wing and you find her outside, you will know the hive has tried to swarm. The queen would have tried to leave with the swarm but could not fly (see swarm control techniques in Unit 15). If you are not de-queening when you find the ball of bees outside the hive, brush them onto a piece of cardboard and shake them on the entrance so the queen can go back into the hive and go to work. If you are de-queening, you have found your queen.

Dumping the Bees to De-Queen the Hive

Equipment needed:

- [ ] Hive tool
- [ ] Smoker or hive bomb
- [ ] Frame grip
- [ ] Bee brush
- [ ] Empty super
- [ ] Pair of 18" two-by-fours
- [ ] Small jar with lid or a queen catcher

Step-by-step directions:

- ■ **Dress properly.**
- ■ **Light the smoker** or have hive bomb handy.
- ■ **Enter the hive.** Place the inverted hive top on the ground nearby. Place the pair of two-by-fours about twelve inches apart on the ground.
- [ ] Set up a ramp to the hive entrance. Place the bottom end of the ramp on a bedsheet or on flat cardboard if the hive is in a grassy area.
- [ ] Using the hive tool, loosen, lift, and smoke between all honey supers (if any) and between the brood chambers. Wait one or two minutes. Set the honey supers off onto the two-by-fours.

- ☐ Set the upper brood chamber on the inverted hive top. Do not remove the queen excluder. Set the empty super on the queen excluder that is still on the upper brood chamber.
- ☐ *Begin dumping with the lower brood chamber.* You will not expect to find the queen in the lower brood chamber, but you never can tell. Usually in a two-story brood system, most of the brood rearing takes place in the upper brood chamber. If you start with the upper brood chamber, you have to dump all the bees out in front of the hive a second time, because they will be running into the lower brood chamber.
- ☐ Remove frames 1 and 2 from the *lower* brood chamber, bees and all. Place them in the empty super temporarily. With the hive tool, loosen all the remaining frames in the *lower* brood chamber, so when you begin to shake the bees from the frames you can do them one after the other as quickly as possible. This way you can watch for the queen running for the hive entrance, just in case she decides to run, and the bees will not be able to form the ball around her.
- ☐ Remove frames 1 and 2 from the empty super, one at a time, and shake the bees off the frames at the entrance to about three feet in front of the hive. These bees will lead the run back into the hive.
- ☐ One by one, as quickly but as carefully as possible, remove and shake all the frames from the lower brood chamber in a pile about three feet in front of the hive. Store the shaken frames in the empty super, or stand them in a safe place nearby.
- ☐ Quickly remove the now-empty lower brood chamber from the bottom board, and brush or bump the super on the ground to remove the bees from inside the super. Brush the bees from the bottom board out onto the ramp. Place the lower brood chamber back on the bottom board and replace all the frames, preferably in the same order as they were removed.
- ☐ Stand by and watch the bees as they run back into the hive. You won't expect to see the queen running for the entrance, but with bees, expect the unexpected. If you see the queen, herd her into the jar and cover it, or place the queen catcher over the queen and close it around her, then remove the queen from the area.
- ☐ After about five minutes or so, most of the bees will have run into or onto the outside of the lower brood chamber, and you will begin to see a number of piles of bees on the ground in the area where you dumped the bees. One of these piles (or none, if the queen is in the upper brood chamber and has not been dumped yet) may contain the queen. The other piles are probably very young bees gathered together in the confusion of being dumped out of the hive.
- ☐ With a gloved or ungloved finger, tip of a brush, or a twig, stir each pile gently, and the bees will begin to move. When you stir the pile

with the queen, you will see the queen begin to move as the bees in the pile begin to move. Herd the queen into the jar, or close the queen catcher over her and remove her from the area.

☐ If none of the piles of bees from the *lower* brood chamber has the queen, then you must dump the *upper* brood chamber in front of the hive, as previously directed. In a two-brood chamber hive, expect to find the queen in the upper brood chamber. (See Figure 16-4.)

☐ When you have found the queen, replace all frames in the upper and lower brood chambers, set the upper brood chamber on the lower brood chamber, and re-queen as directed in the following section.

RE-QUEEN THE HIVE

Equipment needed:

☐ Hive tool
☐ Smoker or hive bomb
☐ Frame grip
☐ New, mated queen, caged
☐ Toothpick or small nail
☐ Empty super

Step-by-step directions:

If you have just finished de-queening the hive, bring out your new caged queen and skip the next five steps. Start with the step that begins "If your queen cage is wooden. . . ."

■ **Dress properly.**
■ **Light the smoker** or have hive bomb handy.
■ **Enter the hive.** Place the inverted hive top on the ground nearby.
☐ With the hive tool, loosen, lift, and smoke between all the honey supers on the hive, if any. Wait one or two minutes, then set the honey supers off onto the inverted hive top.
☐ Smoke the frame tops of the upper brood chamber and gently remove the queen excluder. Set it aside temporarily.
☐ If your queen cage is wooden and contains a candy plug on one end, follow the steps starting with *A*. If your queen cage is a thin cardboard box, follow the steps starting with *B*.

A. *Wooden cage:* With the hive tool between frames 5 and 6, pry the frames apart at both ends. You are trying to make space for the queen cage between the frames, if possible, or to get as much space as possible so the bees can have room to feed and touch the new queen through the screen of the cage.

RE-QUEENING THE HIVE 89

☐ Check to make sure your queen is alive and moving. You must decide whether or not to perforate the candy plug. Use the following as a guide.
1. In early spring or after the honey flow, do not perforate the candy plug. *Reason:* When the flow is light, the bees may release the queen too quickly, before she has been properly introduced. (You will need about four days for introduction.)
2. During mid season re-queening, perforate the candy plug. *Reason:* During the honey flow, the bees may not be as active at eating away the candy as they would if no nectar were coming into the hive. The queen may not be released for a week or more. Perforating the candy plug lets the bees release the queen sooner. (See Figure 13-1.)
3. Alternative: Do not remove the covering or perforate the candy plug, but plan to release the queen manually on the fourth day. Then you will know that she has been introduced for the full four days.

☐ For item 1: Remove the cork plug on the candy end and go to

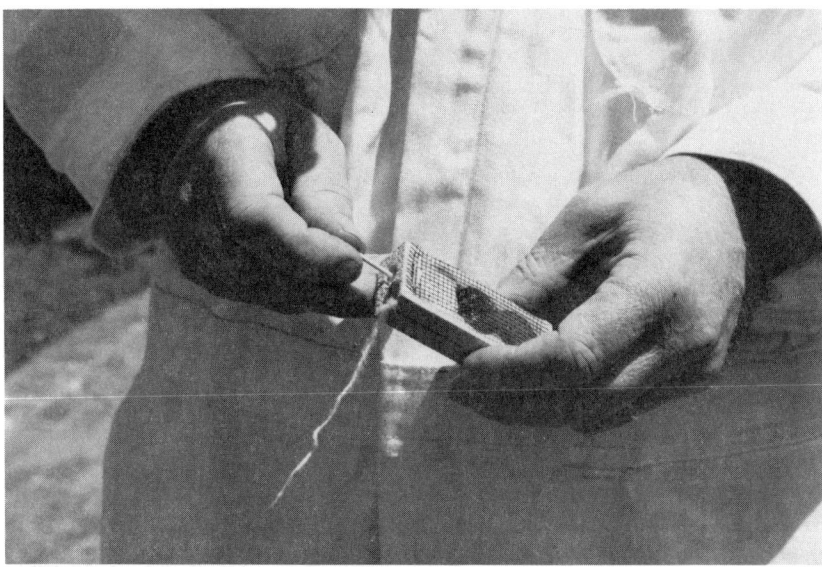

Figure 13-1, Perforating the Candy Plug in the Queen Cage. The queen cage is ready to be hung from a frame top or set on top of the brood frames as soon as the candy plug is perforated. Notice the placement of the thumbtack and string on the end of this queen cage. This placement will not interfere with the bees eating away the candy or the queen exiting from the cage.

the next step. For item 2: Using a toothpick or small nail, perforate the candy plug and go to the next step. For item 3: Do not remove the cork from the candy end of the queen cage, but go to the next step. *Do not release the queen.*

☐ If the space between frames 5 and 6 is wide enough to accept the queen cage, slip it between the frames, screen side down, and push the frames against the queen cage to hold it securely.

If the space between frames 5 and 6 is *not* wide enough to accept the queen cage, set the queen cage lengthwise with screen side down along the space between the frames. The bees must be able to feed and touch the queen for introduction to be accomplished. (See Figure 13-2.)

☐ Replace the queen excluder and honey supers (if any), and close the hive. If there are no honey supers, replace the inner cover *rim side down,* and close the hive.

☐ Wait four days. After that time, check to see if the queen has been released, or release the queen manually. Go to the section entitled "After Four Days."

B. *Cardboard box:* Remove frames 1 and 2. Place them in the empty super.

☐ With the hive tool, loosen frames 3 and 4 and push them to the side wall of the brood chamber. Loosen and remove frame 5.

Figure 13-2, Introducing the Queen Cage on the Brood Frame Tops. Whether or not you perforate the candy plug, set the cage on the frame tops, straddling the enlarged opening made by forcing the brood frames to each side of the brood super with the hive tool.

(You may want to shake the bees from the frame back into the brood chamber.) The frame into and onto which you will push the queen cage should have covered brood, if you have any. If you cannot find any covered brood or do not have any, proceed with the next step.

- [ ] Place the queen cage against an area of covered brood, if you have any; otherwise, place it against the empty brood comb cells. Slide the cover of the box aside while holding the open side of the inner part of the box against the comb. When the cover of the box is removed, press the inner part of the box well into either the covered or empty brood comb. Make sure the box will not fall from the comb before you release it completely. To be sure the box will not fall from the comb, place a large rubber band around the frame to hold the box against the comb. The bees will tear away the cardboard box to release the queen over the next few days. (See Figure 13-3.)
- [ ] Replace the frame containing the queen cage inner box, being careful not to knock the box loose. Slide frames 3 and 4 back into place, and replace frames 1 and 2. Replace the queen excluder and honey supers (if any). Close the hive.
- [ ] *Wait four days.* If you cannot get back to the hive in four days to make sure the queen has been released, do not worry, for the bees will completely tear away the box in time. It would be a good idea to check after four days if you have used the rubber band suggested above, because you should remove it and any remains of the box.

After Four Days

Equipment needed:

- [ ] Hive tool
- [ ] Smoker or hive bomb
- [ ] Frame grip
- [ ] Empty super

Step-by-step directions:

- ■ **Dress properly.**
- ■ **Light the smoker** or have hive bomb handy.
- ■ **Enter the hive.** Place the inverted hive top on the ground nearby.
- [ ] With the hive tool, loosen, lift, and smoke between the honey supers (if any). Wait one or two minutes, then set the honey supers aside on the inverted hive top.

92   KEEPING BEES

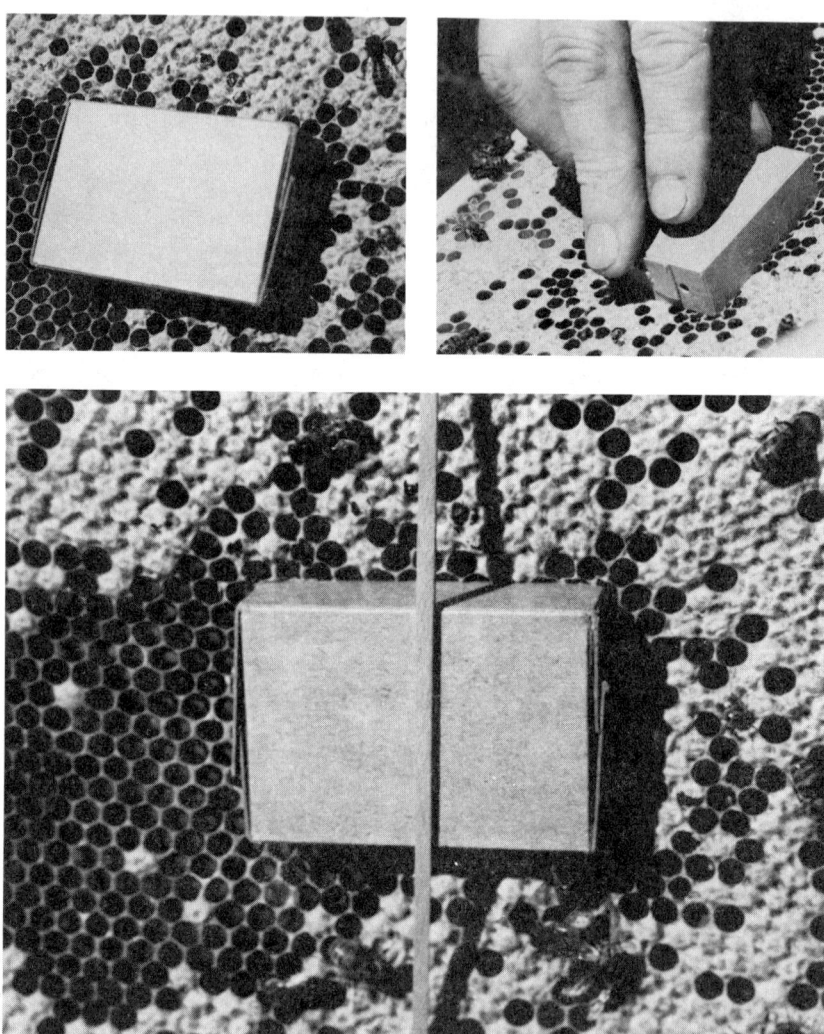

Figure 13-3, Introducing a Queen in the Cardboard Cage. This series of photos shows how to place the inner cardboard box containing the queen into the brood comb. Shake the frame of brood free of most bees, carry the frame from the area (optional), remove the gloves, and slide the outer box off the inner box (See Figure 4-1) while holding the inner box against the brood comb so the queen will not escape. Press the inner box well into the brood comb so that it will hang by itself, or you can attach it with a large rubber band (as shown here) or with string to be sure it will not fall off. Return the frame with the attached queen box as soon as possible, being careful not to knock the box loose.

A. *Wooden cage:* Smoke the frame tops of the upper brood chamber, and gently remove the queen excluder or inner cover. Set it aside temporarily.

- [ ] Remove the queen cage from the top bars of the upper brood chamber. Check to see that the candy plug has been eaten away and that the queen is not dead in the queen cage. (This happens only occasionally. There is more chance of the queen dying in the cage if you have touched the queen in releasing the attendants before you introduced the queen.)
  1. If the queen is not in the cage, close the hive after you replace the queen excluder and honey supers (if any). *Wait seven days* before you inspect the hive to see if the queen is laying.
  2. If the queen is still in the cage, follow the next three steps.
- [ ] If the queen is still in the cage, remove the cork from the opposite end of the queen cage. This is the end that permits the queen direct exit into the hive as soon as she finds it. *Do not* attempt to shake the queen from the cage, for she may become excited and fly away, never to be seen again. Just let her take her own time leaving the cage. Place the queen cage back on the frame tops and close the hive, but *do not* replace the queen excluder or the honey supers (if any).
- [ ] Wait thirty minutes and reenter the hive; check to see that the queen has exited the queen cage. If she has not left, replace the cage, close the hive, and wait another thirty minutes. When the queen has exited the cage, remove the cage, replace the queen excluder and honey supers (if any), and close the hive.

  If robbing might occur while you are waiting for the queen to exit the cage, move any honey supers to a closed area, or take other precautions to insure that robber bees will not get to the honey in the supers.
- [ ] *Wait seven days.* Go to the section entitled "After Seven Days."

B. *Cardboard box:* Remove frames 1 and 2. Place them in the empty super.

- [ ] Loosen and move frames 3 and 4 to the side wall of the brood chamber.
- [ ] Loosen, center, and remove frame 5. Gently remove the remnants of the cardboard box. If it appears that the box has not been torn away enough for the queen to have moved out, set the frame down in the brood chamber before gently removing (lifting off) the box. Cut or remove the rubber band if one was used to hold the box in place.

- [ ] Replace all frames in the brood chamber, replace the queen excluder and honey supers (if any), and close the hive.
- [ ] *Wait seven days* before inspecting the brood chamber.

After Seven Days

Equipment needed:

- [ ] Hive tool
- [ ] Smoker or hive bomb
- [ ] Frame grip
- [ ] Empty super

Step-by-step directions:

- ■ **Dress properly.**
- ■ **Light the smoker** or have hive bomb handy.
- ■ **Enter the hive.** Place the inverted hive top on the ground nearby.
- [ ] With the hive tool, loosen, lift, and smoke between the honey supers (if any) and the upper and lower brood chambers. Wait one or two minutes, then set the honey supers on the inverted hive top nearby.
- [ ] Smoke the frame tops of the upper brood chamber, and gently remove the queen excluder (if any). Set it aside nearby.
- ■ **Inspect the brood chambers.** (See Unit 12.) You will be looking for eggs and tiny larvae. Check that there are no multiple eggs in any cells. If you find no eggs or small larvae in any of the brood cells, there is a better-than-average chance your queen has died. Wait one more week; if you still find no eggs in the brood frames, order another mated queen as soon as possible. When your new queen arrives, follow the steps to re-queen the hive.

Unit 14

Installing a Package of Bees

WHEN THE PACKAGE OF BEES ARRIVES

The package in which the bees are shipped will be either a lightweight wooden box or a sturdy cardboard box screened on two sides. The cardboard box seems to be used by most package bee producers these days. (See Figure 3-1.)

When you get your package of bees, check that there are not an exceptional number of dead bees on the bottom of the package. The wooden package is easier to check. The cardboard package has a smaller screen area, and you may need to tip the box to check the bottom of the package. *Remember* that most producers add about one-half pound of extra bees in the two- and three-pound packages to allow for some bees dying in transit.

If it appears that there is an exceptional death loss when you get your package, make this condition known to the delivery person, and obtain a slip confirming that the package arrived in poor shape. If the package was insured, make a claim for the loss, and order and pay for another package of bees. If the producer guaranteed live delivery (if the package was not insured), send the producer a copy of the statement given you by the carrier, and he or she will send you another package. Most producers now include insurance in the cost of the package.

Unless the carrier wants to keep the package of bees (if it is insured, he or she may), accept the package of bees and your statement that they arrived in poor condition, if in fact they did. You are the best hope for the package of bees that arrives in poor condition. Most carriers do not know what to do with an unwanted package of bees, and the producer does not want to pay postage to get the unwanted package back. You can try to keep those bees alive until the new package arrives by either putting them in the hive and feeding them or adding the bees to an existing hive. If the queen is still alive, you may be able to use her to re-queen one of your hives.

FEEDING THE PACKAGE OF BEES WHEN YOU GET THEM HOME

You can prepare the sugar solution to feed the bees several days before you expect them to arrive. You can also buy or clean up a trigger-actuated mist sprayer (such as many household cleaners are now packaged in and sold). You may want to make some extra sugar solution to feed the bees for several weeks after they have been installed in their new hive. You can use the entrance feeder to feed the sugar solution over the long term.

Preparing the Sugar Solution to Feed the Bees

You can make the sugar solution as directed in the following steps for feeding the package now, or you can make a gallon of sugar solution at this time. (See Unit 11.)

- [ ] Place one quart of water in a two-quart pan and heat to boiling. Turn off the heat.
- [ ] Add one quart of sugar to the pan of hot water. Stir to dissolve the sugar. Let the solution cool to 90° F before feeding.
- [ ] Fill a quart jar with the sugar solution. You will use this solution to feed the package when you install it in the hive.
- [ ] Place the remaining sugar solution in the trigger mist sprayer to use in feeding the bees now and for gorging the bees later, just before installing them in the hive. If you need more sugar solution for the trigger sprayer, take some from the quart jar.

Feeding the Package of Bees

- [ ] When the sugar solution has cooled sufficiently (90°F or less), pick up the package of bees and mist spray five to seven pumps of solution through the screen on one side of the package. Turn the package around, and repeat the spraying on that side of the package. (See Figure 14-1.)
- [ ] Place the package in a cool spot until about one hour before installing the package of bees in the hive. Before installing the bees in the hive, you will gorge the bees as described later.

HOLDING A PACKAGE OF BEES FOR SEVERAL DAYS, IF NECESSARY

If for some reason you cannot install the package of bees for several days, you can hold them by performing the previous two steps each

INSTALLING A PACKAGE OF BEES 97

Figure 14-1, Feeding or Gorging the Package of Bees. A trigger mist sprayer is being used to feed the package of bees through the side screen.

morning and evening until the bees can be installed in the hive. Keep the bees in a cool place while holding them.

It is best to install the package of bees as soon after receiving them as possible. The bees can be installed at any time of the day, but about one hour before dark is best because they must settle in for the night, and there is less chance they might abscond (which seldom happens).

GORGING THE BEES BEFORE INSTALLING THEM IN THE HIVE

There are several reasons for gorging the bees before installing the package in the hive.

1. A bee gorged with honey/nectar/ sugar solution is bound to have a gentle disposition. Gorging makes the bees easier to handle and tends to settle them down on the comb or foundation after they are installed.
2. Bees whose stomachs are full of honey/nectar/sugar solution and have no place to store it (such as on foundation) hang around, and their wax glands begin to secrete wax flakes that they can turn into comb. The quicker the comb is drawn, the sooner the queen can begin to lay. The reason you feed sugar solution for several weeks after installing the bees is to keep the comb building moving along rapidly.

Step-by-step directions:

☐ Begin to gorge the bees about one hour before installing the package in the hive.

- [ ] Lift the package and spray fifteen to twenty pumps of mist sugar solution onto the bees through the screen. Do not use so much that it runs and drops off the bees. (See Figure 14-1.)
- [ ] Turn the package around and repeat the spraying on that side.
- [ ] Set the package aside for about ten minutes while the bees clean up the sugar solution that was sprayed on them.
- [ ] Repeat the spraying and set the package aside. Continue spraying and setting the package aside. As the bees slow down in clean-up, drop the package down on the floor from about one foot to knock the bees from their cluster around the feeder can in the package. As the bees recluster, continue to spray and set aside. You will probably spray and set aside about five to six times.

REMOVING THE QUEEN CAGE FROM THE PACKAGE OF BEES

When you are ready to install the package in your hive, remove the cover board or cardboard cover from the package using your hive tool, as directed. Bees will not escape as the cover of the package is removed. A feeder can blocks the exit of the bees until it is removed.

If your bees arrive in a wooden package, when the cover is removed you will see the bottom of the feeder can and, in a slot beside the can, either a wire connected to a tack or a plastic knot or star. The wire or plastic strip is connected to the queen cage, which is hanging below, next to the feeder can.

If your bees arrive in the cardboard package, when the cover is removed you will see the bottom of the feeder can and, looking around the circumference of the feeder can, a strip of aluminum folded over the can or the cardboard. The queen cage is attached to the strip of metal and is hanging below, next to the feeder can.

Do not remove the queen cage from the package until you are installing the package of bees into the hive. Then when you are directed, remove the queen cage from the package by following these steps.

- [ ] Remove the cover from the package, and determine the queen-holding arrangement as mentioned previously.
- [ ] Lift the package about one foot and drop it down squarely on the ground to knock the cluster of bees from the can and the queen cage.
- [ ] *Wooden package:*
    1. Loosen the wire from the tack, hold the wire, and lift out the feeder can. Slide the wire out of the slot and lift out the queen cage. Replace the feeder can to keep the bees in the package.

2. Grasp the plastic knot or cross, lift out the feeder can, slide the plastic strip out of the slot, and lift out the queen cage. Replace the feeder can to keep the bees in the package.

☐ *Cardboard package:* Grasp the metal strip, lift out the feeder can, lift out the queen cage, and replace the feeder can to keep the bees in the package.

Queen Cages in Package Bees

Two types of queen cages are in general use for shipping with package bees. (See Figure 3-1.)

1. The wooden queen cage may or may not contain a candy plug. In most cases, the wooden queen cage shipped with a package of bees will not have a candy plug and seldom has any attendants with the queen, because the bees in the package will feed and tend the queen. If the cage contains candy, it provides for the automatic release of the queen when the cork is removed, and the bees eat away the candy. If there is no candy in the cage, you will have to release the queen manually after four days, as was explained in the Unit 13.
2. The plastic queen cage never contains candy, and the queen always must be released manually. The plastic queen cage has perforations on the front and two sides so the bees can tend the queen in transit and during introduction. On the top and bottom of the plastic cage are openings, each covered with a plastic cap. After four days of introduction, one or both of these caps can be removed for the queen to join the bees and begin to lay.

METHODS FOR INSTALLING THE PACKAGE OF BEES

Installing a package of bees in a hive is not really difficult. If you get the queen cage and the bees in the hive, it is only very seldom that the bees will not settle in and go to work. A problem might occur if you are installing three or four packages, one after the other, in adjacent hives in the apiary. In this case there may be considerable drifting of the orienting bees as they settle in. If this is of some concern to you and you have the time, install one package and wait an hour or two, then install the next package, etc., until all packages are installed. This gives the bees in each package time to settle in before the next package is installed, and those bees begin to fly in elimination and orientation flights. You may install the package on foundation or on comb, either frames of comb from last year's extracted honey or on the combs of a hive that has died out for any reason except American Foulbrood Disease.

In the North: If you have found a disease-free hive that died out over winter, you will not need to store it under PDB moth crystals if you have ordered a package of bees to restock it. Until you receive your package, move the hive to a bee-free area, or close the hive completely so other bees cannot steal any remaining honey.

In the South: If you have found a disease-free dead hive (at any time of the year), store the brood supers in plastic garbage bags with PDB moth crystals as soon as possible. Be sure to air the supers for two to three days before installing a package (or a swarm) on those frames.

Order the three-pound package of bees for starting a new hive on frames of foundation. A two-pound package will do to start the new hive on comb or brood comb from a disease-free dead hive. In every case, a three-pound package will bring the hive along faster.

When installing a package of bees on foundation or comb, it makes good sense to *medicate*, particularly if the hive is installed on comb from a dead hive. Medicate to control AFB/EFB and Nosema Disease, as described in Unit 11.

Adding a pollen supplement patty after installing the package can help the hive move along faster in the brood-rearing cycle, particularly in the North, because it is a food source that will be readily available to the bees until natural pollen is available in the field. Feeding the sugar solution along with the pollen supplement provides all the food requirements for brood rearing to begin, particularly if rainy or cold weather sets in after the hive is started.

There could be between 3500 and 5000 bees per pound, so you can see that any help in food supplies as the hive is getting started would be welcomed. The first new replacement bee emerges twenty-one days after the queen is released and begins to lay. The most critical time for the newly installed package of bees is about three weeks after the bees have been installed, because the forty-five day life span of the older bees in the package is running out. If brood rearing begins soon after the package is installed, the population begins to build up quickly.

Three methods for installing the package of bees in the hive are described here. Read through all the methods, and chose the one you like.

Method 1

Have everything you need on hand at the hive, which should be standing on its permanent location. Having gorged the bees within the

INSTALLING A PACKAGE OF BEES 101

last hour, as instructed earlier in this unit, you are ready to begin. A smoker will not be needed. (See Figure 14-2.)

Equipment needed:

- ☐ Hive tool
- ☐ Package of bees
- ☐ One foot of fine wire or string
- ☐ Thumbtacks (2-4)
- ☐ Toothpick or small nail
- ☐ Empty medium or deep super
- ☐ Feeder base and quart jar of sugar solution
- ☐ Scrap block of wood, about 8" long

Step-by-step directions:

■ Dress properly.

- ☐ Remove the top and inner cover from the hive. Set these aside.
- ☐ Remove frames 5 and 6 from one side of the brood super, and stand them in a safe place nearby or place them in the empty super.
- ☐ With the hive tool, remove the cover from the package of bees. The bees will not escape when the cover is removed. Check the queen-

Figure 14-2, Method 1-- Installing a Package of Bees. The package of bees is laid on its side on the frame tops of the brood chamber, inside the empty super. The opening of the package should be near the frame on which the queen is hanging. Notice that the feeder can is also placed on the frame tops inside the empty super.

holding arrangement under the cover and reread the section in this unit entitled "Removing the Queen Cage from the Package of Bees."
- [ ] Lift the package about one foot and drop it down squarely on the ground to knock the bees from the feeder can and the queen cage.
- [ ] Remove the queen cage from the package, and replace the can or cover to keep the bees in the package. Some bees will escape, but do not worry about them.
- [ ] Check the queen cage to make sure the queen is alive and moving. If she is dead (which seldom happens), continue with the following steps. When you have finished installing the package, call (preferably) or write the producer for a new queen if your queen was dead. *Note:* Install the queen as directed, even if she is dead. This will keep your bees from absconding until the new queen arrives.
- [ ] If you have a plastic cage or a wooden cage with no candy, go to the next step. If you have a wooden cage with candy, you must decide at this time whether to perforate the candy plug and let the bees release the queen, or proceed as though the cage had no candy and release the queen manually after four days. As mentioned earlier, the bees sometimes release the queen too quickly (if you re-queen) before or after the honey flow. If you plan to let the bees release the queen, remove the closure and/or cork on the candy end of the queen cage, and perforate the candy plug with a toothpick or small nail. (See Figure 13-1.)
- [ ] Remove frame 7 from one side of the super (counting any frames removed earlier). Place a thumbtack in the center of that frame top.
- [ ] If your queen cage is wooden, contains candy, and you have perforated the candy, place a thumbtack on the end of the queen cage containing the candy you have just perforated. The tack must not obstruct the opening on the end of the cage, for the bees must be able to eat away the candy to release the queen. Fasten the wire or string to the tack on the end of the queen cage, then hang the cage about two inches below the frame top from the thumbtack in the center of the frame top. Be sure the screen of the cage is facing away from the foundation or comb so the bees can feed and touch the queen. *Skip* the next two steps.
- [ ] If your queen cage is wooden and contains no candy, or if you have decided not to perforate the candy plug in a queen cage with candy, place a thumbtack on either end of the cage in such a way that the tack does not interfere with the opening on that end. (The cork must be removed in four days, and the queen must be able to move out of the opening.) Fasten the wire or string to the tack on the end of the cage, then hang it about two inches below the frame top by the thumbtack in the center of the frame top. Be sure the screen

INSTALLING A PACKAGE OF BEES   103

of the cage is facing away from the foundation or comb so the bees can feed and touch the queen. If your cage has the aluminum strip and it is long enough, you can remove the tack in the center of the frame top and push the tack back in place through the aluminum strip. *Skip* the next step.

☐ If your queen cage is plastic, fasten the wire or string to the plastic strip attached to the queen cage. Hang the plastic cage two inches below the frame top by the thumbtack in the center of the frame top. Be sure the perforations of the cage face away from the foundation or comb so the bees can touch and feed the queen.

Note: If you are using comb or brood comb from an old hive, it may be necessary to remove one of the frames for the next four days in order for the queen cage to hang so the bees can feed and touch the queen. Check that there will be room between the combs, and remove a frame, if necessary.

☐ Replace the frame on which the queen cage is attached. Check to be sure the cage is hanging properly—screen or perforations facing away from the foundation or comb. (See Figure 14-3.)

☐ If the package is being installed on foundation, replace the two frames removed earlier. If the package is being installed on comb, replace one of the frames and store the other frame of comb until

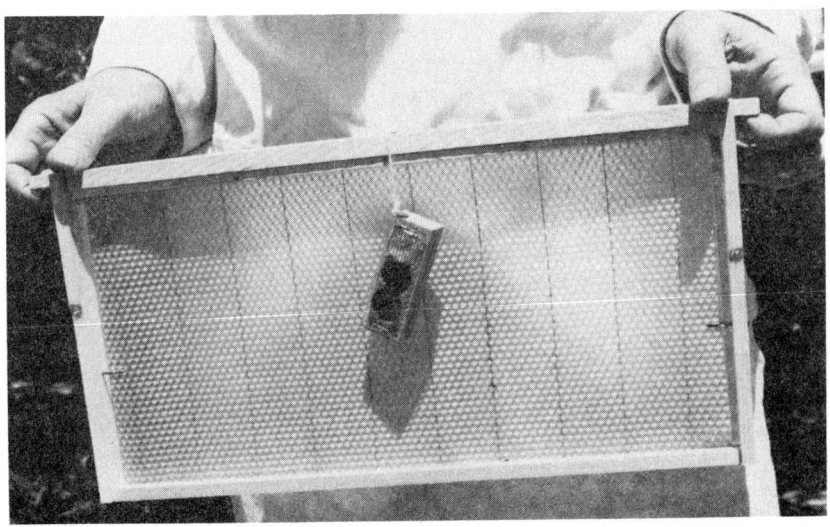

Figure 14-3, Queen Cage Hanging from a Frame Top. The queen cage is hanging from the frame top so that the bees can release the queen or so that she can be manually released after four days. The queen in package bees can also be introduced on the brood frame tops. (See Figure 13-2.)

the four-day inspection. The reason for leaving out one frame of comb is to make sure there is space around the queen cage for the bees to become properly introduced while feeding and touching the queen.
- [ ] Place the empty super on top of the brood chamber in which the queen cage was placed.
- [ ] Lift the package one foot and drop it down squarely on the ground. Remove the feeder can if you replaced it in the package earlier, or remove the cover and set the package on its side, inside the empty super on the frame tops of the brood chamber. The opening of the package should be about where the queen cage is tied below. The feeder can, if it contains sugar syrup, also can be set on the frame tops inside the empty super. If you are using a deep super for the empty super and comb on which to install the package, you can store the extra frame of comb in the super as well. It will be safe and available to replace in the hive when the queen cage is removed in four days.
- [ ] Replace the inner cover and hive top.
- [ ] Insert the feeder base in the hive entrance at one side. Invert into the feeder base the quart of sugar solution, with the feeder cap screwed on securely.
- [ ] Close the hive entrance down to about three inches alongside the feeder with the scrap block of wood. There may be many bees flying about at this time. They will return to the hive and will all be inside by dark.
- [ ] *Do not enter the hive for four days.*
- [ ] Keep the feeder jar filled. *Caution:* Be careful when removing the feeder cap from the jar. The edges of the cap are quite sharp, and you could get a bad cut. Use gloves or a rag to remove the feeder cap.
- [ ] See the section in this unit entitled "After Four Days" to continue the manipulative steps on installing the package of bees in the hive.

Method 2

Have everything you need on hand at the hive, which is standing in its permanent location. Having gorged the bees within the last hour, as instructed earlier in this unit, you are ready to begin. The smoker will not be needed.

Equipment needed:

- [ ] Hive tool
- [ ] Package of bees

- [ ] One foot of fine wire or string
- [ ] Toothpick or small nail
- [ ] Thumbtacks (2-4)
- [ ] Feeder base and quart jar of sugar solution
- [ ] Scrap block of wood, about 8" long

Step-by-step directions:

■ **Dress properly.**
- [ ] Remove the inner cover and hive top and set them aside.
- [ ] Remove the first five frames from one side of the brood chamber. Take them to the house and store them safely, for they will be out of the hive for the next four days.
  Note: If your package is cardboard, you may only have to remove four frames. You can check by setting the package in the space when four frames have been removed; if the package fits, then remove only four frames. (See Figure 14-4.)
- [ ] With the hive tool, remove the cover from the package of bees. Bees will not escape when the cover is removed. Check the queen-holding arrangement and reread the section in this unit entitled "Removing the Queen Cage from the Package of Bees."
- [ ] Lift the package about one foot and drop it down squarely on the ground to knock the bees from the feeder can and the queen cage.
- [ ] Remove the queen cage from the package, and either replace the

Figure 14-4, Method 2-- Installing a Package of Bees. When the queen cage has been removed and tied to the frame top after removing four or five frames, the package is set in the brood chamber. The feeder can is removed, and the inner cover and hive top are placed on the hive.

feeder can or place the cover over the package to keep the bees in the package. Some bees will escape, but do not worry about them.

☐ Check the queen cage to make sure the queen is alive and moving. If she is dead (which seldom happens), continue with the following steps. When you have finished installing the package, call (preferably) or write the producer for a new queen, if your queen was dead. *Note:* Install the queen as directed, even if she is dead. This will keep your bees from absconding until the new queen arrives.

☐ If you have a plastic cage or a wooden cage with no candy, go to the next step. If you have a wooden cage with candy, you must decide at this time whether to perforate the candy plug and let the bees release the queen, or proceed as though the cage had no candy and release the queen manually after four days. As mentioned earlier, the bees sometimes release the queen too quickly (if you re-queen) before or after the honey flow. If you plan to let the bees release the queen, remove the closure and/or cork on the candy end of the queen cage, and perforate the candy plug with a toothpick or small nail. (See Figure 13-1.)

☐ Lift out frame 5 or 6, depending on how many frames you stored away earlier. Place a thumbtack in the center of that frame top.

☐ If your queen cage is wooden, contains candy, and you have perforated the candy, place a thumbtack on the end of the queen cage containing the candy you have just perforated. The tack must not obstruct the opening on the end of the cage, for the bees must be able to eat away the candy to release the queen. Fasten the wire or string to the tack on the end of the queen cage, then hang the cage about two inches below the frame top from the thumbtack in the center of the frame top. Be sure the screen of the cage is facing away from the foundation or comb so the bees can feed and touch the queen. *Skip* the next two steps.

☐ If your queen cage is wooden and contains no candy, or if you have decided not to perforate the candy plug in a queen cage with candy, place a thumbtack on either end of the cage in such a way that the tack does not interfere with the opening on that end. (The cork must be removed in four days, and the queen must be able to move out of the opening.) Fasten the wire or string to the tack on the end of the cage, then hang it about two inches below the frame top by the thumbtack in the center of the frame top. Be sure the screen of the cage is facing away from the foundation or comb so the bees can feed and touch the queen. If your cage has the aluminum strip and it is long enough, you can remove the tack in the center of the frame top and push the tack back in place through the aluminum strip. *Skip* the next step.

- [ ] If your queen cage is plastic, fasten the wire or string to the plastic strip attached to the queen cage. Hang the plastic cage two inches below the frame top by the thumbtack in the center of the frame top. Be sure the perforations of the cage face away from the foundation or comb so the bees can touch and feed the queen.

  Note: If you are using comb or brood comb from an old hive, it may be necessary to remove one of the frames for the next four days in order for the queen cage to hang so the bees can feed and touch the queen. Check that there will be room between the combs, and remove a frame, if necessary.

- [ ] Replace the frame on which the queen cage is attached. Check to be sure the cage is hanging properly—screen or perforations facing away from the foundation or comb. (See Figure 14-3).
- [ ] Lift the package one foot and drop it down squarely on the ground. Place the entire package in the empty space in the brood chamber.
- [ ] Lift out the feeder can or remove the cover from the package. Slide the inner cover onto the hive, and replace the hive top. If the feeder can contains sugar solution, set it on top of the hive and pour it into your feeder jar when you refill it later.
- [ ] Insert the feeder base in the hive entrance at one side. Invert into the feeder base the quart jar of sugar solution, with the feeder cap screwed on securely.
- [ ] Close the hive entrance to an opening of about three inches alongside the feeder with a scrap block of wood. Do not be concerned about any flying bees, for they will all return to the hive by dark.
- [ ] *Do not enter the hive for four days.*
- [ ] Keep the feeder jar filled. *Caution:* Be careful when removing the feeder cap from the jar. The edges of the cap are quite sharp, and you could get a bad cut. Use gloves or a rag to remove the feeder cap.
- [ ] See the section in this unit entitled "After Four Days" for the next manipulative steps on installing the package of bees in the hive.

Method 3

This method should be used only if the temperature is 70° F or more at the time the package is installed. Have everything you need on hand at the hive, which is setting in its permanent location. Having gorged the bees within the last hour, as instructed earlier in this unit, you are ready to begin. A smoker will not be needed.

Equipment needed:

- [ ] Hive tool

- [ ] Package of bees
- [ ] One foot of fine wire or string
- [ ] Thumbtacks (2-4)
- [ ] Toothpick or small nail
- [ ] Hose attached to a faucet nearby
- [ ] Feeder base and quart jar of sugar solution
- [ ] Scrap block of wood, about 8" long

Step-by-step directions:

■ **Dress properly.**
- [ ] Remove the inner cover and hive top, and set them aside.
- [ ] Remove frames 3, 4, and 5 from one side of the brood chamber. Stand them on end in a safe place nearby.
- [ ] Pick up the package and move some distance from the hive. Turn on the faucet until a gentle stream comes from the hose.
- [ ] Pick up the package, turn it on its side, and run water over and through the bees clustered in the package. Turn the package over and continue wetting the bees from that side. Run several quarts of water over and through the bees. (See Figure 14-5.)
- [ ] Turn off the faucet and return with the package to the hive. Make sure the package leaks enough so that it will not hold water. You only want to wet the bees, not drown them.
- [ ] With the hive tool, remove the cover from the package of bees. The bees will not escape when the cover is removed. Check the queen-holding arrangement under the cover and reread the section in this unit entitled "Removing the Queen Cage from the Package of Bees."
- [ ] Lift the package about one foot and drop it down squarely on the ground to knock the bees from the feeder can and the queen cage.
- [ ] Remove the queen cage from the package, and replace the can or cover to keep the bees in the package. Some bees may have escaped, but do not worry about them.
- [ ] Check the queen cage to make sure the queen is alive and moving. If she is dead (which seldom happens), continue with the following steps. When you have finished installing the package, call (preferably) or write the producer for a new queen if your queen was dead. *Note:* Install the queen as directed, even if she is dead. This will keep your bees from absconding until the new queen arrives.
- [ ] If you have a plastic cage or a wooden cage with no candy, go to the next step. If you have a wooden cage with candy, you must decide at this time whether to perforate the candy plug and let the bees release the queen, or proceed as though the cage had no candy and release the queen manually after four days. As mentioned earlier, the bees sometimes release the queen too quickly (if you re-queen)

INSTALLING A PACKAGE OF BEES 109

Figure 14-5, Method 3--
Installing a Package of
Bees. Left: The bees are
being wetted so they cannot
fly readily. Below: The bees
are being dumped on the
frame tops over the queen
cage hanging from frame
top. The inner cover and
hive top will be replaced,
and the bees will settle into
their new home.

before or after the honey flow. If you plan to let the bees release the queen, remove the closure and/or cork on the candy end of the queen cage, and perforate the candy plug with a toothpick or small nail. (See Figure 13-1.)

- [ ] Remove frame 6 from one side of the brood chamber. Place a thumbtack in the center of the frame top.
- [ ] If your queen cage is wooden, contains candy, and you have perforated the candy, place a thumbtack on the end of the queen cage containing the candy you have just perforated. The tack must not obstruct the opening on the end of the cage, for the bees must be able to eat away the candy to release the queen. Fasten the wire or string to the tack on the end of the queen cage, then hang the cage about two inches below the frame top from the thumbtack in the center of the frame top. Be sure the screen of the cage is facing away from the foundation or comb so the bees can feed and touch the queen. *Skip* the next two steps.
- [ ] If your queen cage is wooden and contains no candy, or if you have decided not to perforate the candy plug in the queen cage with candy, place a thumbtack on either end of the cage in such a way that the tack does not interfere with the opening on that end. (The cork must be removed in four days and the queen must be able to move out of the opening.) Fasten the wire or string to the tack on the end of the cage, then hang it about two inches below the frame top by the tack in the center of the frame top. Be sure the screen of the cage is facing away from the foundation or comb so the bees can feed and touch the queen. If your cage has the aluminum strip and it is long enough, you can remove the tack in the center of the frame top and push the tack back in place through the aluminum strip. *Skip* the next step.
- [ ] If your queen cage is plastic, fasten the wire or string to the plastic strip attached to the queen cage. Hang the plastic cage two inches below the frame top by the thumbtack in the center of the frame top. Be sure the perforations of the cage face away from the foundation or comb so the bees can touch and feed the queen.

Note: If you are using comb or brood comb from an old hive, it may be necessary to remove one of the frames for the next four days in order for the queen cage to hang so the bees can feed and touch the queen. Check that there will be room between the combs, and remove a frame, if necessary.

- [ ] Replace the frame on which the queen cage is attached. Check to be sure the cage is hanging properly—screen or perforations facing away from the foundation or comb. (See Figure 14-3.)

- [ ] Lift the package one foot and drop it down squarely on the ground. Remove the feeder can or cover and set it aside.
- [ ] Dump the wet bees out of the hole in the package into the super where the three frames were removed earlier. When most of the wet bees have been shaken and dumped from the package, set the package on its side at the hive entrance. (See Figure 14-5.)
- [ ] Gently replace the three frames in the brood chamber. Replace the inner cover and hive top.
- [ ] Insert the feeder base in the hive entrance on one side. Invert into the feeder base the quart jar of sugar solution, with the feeder cap screwed on securely.
- [ ] Close the hive entrance to an opening of about three inches alongside the feeder with a scrap block of wood.
- [ ] As the bees begin to dry off, many will fly out. Do not worry, for they will all return by dark.
- [ ] *Do not enter the hive for four days.*
- [ ] Keep the feeder jar filled. *Caution:* Be careful when removing the feeder cap from the jar. The edges of the cap are quite sharp, and you could get a bad cut. Use gloves or a rag to remove the feeder cap.
- [ ] See the section in this unit entitled "After Four Days" to complete the manipulatons for installing the package of bees.

About a Dead Queen

If the queen is dead at the time the package is installed, contact the producer by phone (preferably) or letter after you have installed the package. In almost every case, the producer guarantees live delivery of the queen, even if the package is insured, because the package is of very little use or value without a live, laying queen. Be sure to install the dead queen according to the directions given previously. Otherwise, the bees from the package may abscond or join the nearest queen-right hive in your yard.

If the queen is dead in the cage at the four-day inspection, order a new queen as soon as possible. If the queen was alive at the time she was installed, the responsibility for a replacement queen is yours, and you must pay for a new queen. There is a chance that the producer will replace the queen if you call and explain the situation. In any event:

Leave the dead queen in the hive until your new queen arrives.

In either of the aforementioned situations, when the new queen arrives, she must be introduced to the hive for four days by hanging her cage among the bees as you did when installing the original queen.

After Four Days

Four days after installing the package of bees in the hive, check the brood chamber to see if the queen has been released (if you perforated the candy plug) or to release the queen manually from the wooden or plastic cage. *Do not* attempt to release the queen by removing the screen or shaking her out of the cage. The queen is not heavy with eggs at this time and can fly; she may do so if she is excited by the shaking or rough handling of the queen cage. Follow the steps for releasing the queen given on the next few pages.

Remember the method you used for installing the package of bees four days ago, and follow the steps listed for that method. You will want to medicate the bees and add a pollen supplement patty when you make this four-day check of the queen (or release her). Prepare the medications and pollen supplement patties as directed in Unit 11. You can have them ready to go a few days before you make the four-day check.

Method 1

Equipment needed:
- ☐ Hive tool
- ☐ Smoker or hive bomb
- ☐ Frame grip
- ☐ Terramycin®/powdered sugar medication (AFB)
- ☐ One gallon Nosema-medicated sugar solution
- ☐ Pollen supplement patty

Step-by-step directions:
- ■ **Dress properly.**
- ■ **Light the smoker** or have hive bomb handy.
- ■ **Enter the hive.** Remove the empty super from the hive. Remove the package from the frame tops, and set it on its side near the hive entrance. Remove the feeder can from the frame tops.
- ☐ Gently remove the frame (bees and all) that is next to the frame on which the queen cage is hanging. Set the frame in the empty super, or stand it on end nearby.
- ☐ Center the frame on which the queen cage is hanging, gently raise the queen cage, and set it on the frame top. Replace the frame removed in the previous step, and space the ten frames evenly in the brood chamber.

- [ ] Check the queen cage. Has the queen been released? If you perforated the candy plug and the answer is *yes*, remove the queen cage. Place the pollen patty on the frame tops of the brood super, and spoon the AFB medicine (Terramycin®/powdered sugar mix) along the frame tops around the pollen patty. If the answer is *no*, go to the next step.
- [ ] If the queen was not released from a queen cage containing candy, or if the cage contained no candy or was plastic, use one of the following steps for manual release of the queen.
  1. In the wooden cage that contains the candy, remove the cork from the end that does not contain candy (the end that provides the direct exit). Do not let the queen escape.
  2. Remove the cork from either or both ends of the wooden cage that does not contain candy. Do not let the queen escape.
  3. Remove the plastic caps from either or both ends of the plastic queen cage. Do not let the queen escape.
- [ ] Set the queen cage on top of the brood frames, replace the inner cover (deep rim side down), and replace the hive top.
- [ ] Wait fifteen to thirty minutes. Enter the hive gently, and check the queen cage. The queen should be out, but if she is not, replace the inner cover and hive top and wait another fifteen to thirty minutes.
- [ ] When the queen is out of the cage, remove the queen cage, place the pollen patty on the frame tops, and spoon the AFB medication along the frame tops around the pollen patty. Replace the inner cover *(deep rim side down)*, then replace the hive top. (See Figure 11-2.)
- [ ] Begin to medicate to control Nosema Disease at the next filling of the feeder jar. Keep the feeder jar filled until the gallon of Nosema-medicated sugar solution is used up. If you wish to feed the bees after the Nosema-medicated sugar solution is used up, prepare a 1-1 sugar solution as directed in Unit 11.
- [ ] *Do not open the hive for seven days!* See the section in this unit entitled "After Seven Days."

Method 2

Equipment needed:

- [ ] Hive tool
- [ ] Smoker or hive bomb
- [ ] Frame grip
- [ ] Terramycin®/powdered sugar medication (AFB)
- [ ] One gallon Nosema-medicated sugar solution
- [ ] Pollen supplement patty

114   KEEPING BEES

Step-by-step directions:

■ **Dress properly.**
■ **Light the smoker** or have hive bomb handy.
☐ Bring back to the hive the four or five frames you removed and stored when you installed the package of bees. Set them nearby.
■ **Enter the hive.** Lift the package from the hive and set it on its side near the hive entrance. Gently raise the queen cage from where it is hanging and set it on the frame top.
☐ Replace the frames brought back to the hive, and space the frames evenly in the brood chamber.
☐ Check the queen cage. Has the queen been released? If you perforated the candy plug and the answer is *yes*, remove the queen cage. Place the pollen patty on the frame tops of the brood super, and spoon the AFB medicine (Terramycin®/powdered sugar mix) along the frame tops around the pollen patty. If the answer is *no*, go to the next step.
☐ If the queen was not released from a queen cage containing candy, or if the cage contained no candy or was plastic, use one of the following steps for manual release of the queen.

   1. In the queen cage that contains candy, remove the cork from the end that does not contain candy (the end that provides direct exit for the queen). Do not let the queen escape.
   2. Remove the cork from either or both ends of the wooden cage that does not contain candy. Do not let the queen escape.
   3. Remove the plastic caps from either or both ends of the plastic queen cage. Do not let the queen escape.

☐ Set the queen cage on top of the brood frames, replace the inner cover *(deep rim side down)*, and replace the hive top.
☐ Wait fifteen to thirty minutes. Enter the hive gently, and check the queen cage. The queen should be out, but if she is not, replace the inner cover and hive top and wait another fifteen to thirty minutes.
☐ When the queen is out of the cage, remove the queen cage, place the pollen patty on the frame tops, and spoon the AFB medication along the frame tops around the pollen patty. Replace the inner cover (*deep rim side down*), then replace the hive top. (See Figure 11-2.)
☐ Begin to medicate to control Nosema Disease at the next filling of the feeder jar. Keep the feeder jar filled until the gallon of Nosema-medicated sugar solution is used up. If you wish to feed the bees after the Nosema-medicated sugar solution is used up, prepare a 1-1 sugar solution as directed in Unit 11.

☐ *Do not open the hive for seven days!* See the section in this unit entitled "After Seven Days."

Method 3

Equipment needed:

☐ Hive tool
☐ Smoker or hive bomb
☐ Frame grip
☐ Empty deep super
☐ Terramycin®/powdered sugar medication (AFB)
☐ One gallon Nosema-medicated sugar solution
☐ Pollen supplement patty

Step-by-step directions:

■ **Dress properly.**
■ **Light the smoker** or have hive bomb handy.
■ **Enter the hive.** Gently remove the frame (bees and all) that is next to the frame on which the queen is hanging. Set the frame in the empty super, or stand it on end nearby.
☐ Center the frame on which the queen cage is hanging, gently raise the queen cage and set it on the frame top. Replace the frame removed in the previous step, and space the frames evenly in the brood chamber.
☐ Check the queen cage. Has the queen been released? If you perforated the candy plug and the answer is *yes*, remove the queen cage. Place the pollen patty on the frame tops of the brood super, and spoon the AFB medicine (Terramycin®/powdered sugar mix) along the frame tops around the pollen patty. If the answer is *no*, go to the next step.
☐ If the queen was not released from a queen cage containing candy, or if the cage contained no candy or was plastic, use one of the following steps for manual release of the queen.

  1. In the queen cage that contains candy, remove the cork from the end that does not contain candy (the end that allows direct exit for the queen). Do not let the queen escape.
  2. Remove the cork from either or both ends of the wooden cage that does not contain candy. Do not let the queen escape.
  3. Remove the plastic caps from either or both ends of the plastic queen cage. Do not let the queen escape.

☐ Set the queen cage on top of the brood frames, replace the inner cover *(deep rim side down)*, and replace the hive top.

- [ ] Wait fifteen to thirty minutes. Enter the hive gently, and check the queen cage. The queen should be out, but if she is not, replace the inner cover and hive top and wait another fifteen to thirty minutes.
- [ ] When the queen is out of the cage, remove the queen cage, place the pollen patty on the frame tops, and spoon the AFB medication along the frame tops around the pollen patty. Replace the inner cover *(deep rim side down)*, then replace the hive top. (See Figure 11-2.)
- [ ] Begin to medicate to control Nosema Disease at the next filling of the feeder jar. Keep the feeder jar filled until the gallon of Nosema-medicated sugar solution is used up. If you wish to feed the bees after the Nosema-medicated sugar solution is used up, prepare a 1-1 sugar solution as directed in Unit 11.
- [ ] *Do not open the hive for seven days!* See the following section entitled "After Seven Days."

After Seven Days

Note: No matter which method you used for installing the package of bees, follow these steps.

After seven days of uninterrupted activity in the hive, the bees and the queen should be well settled in. It is time to check that the queen is in the hive and that she has begun to lay well. It is always possible that your queen may have been killed by the bees at the time she was released, but this does not happen often. In any case, you will check to see if there are eggs and young larvae in the cells of some frames. You will ■ **Inspect the brood chamber,** following the steps in Unit 12.

FIRST INSPECTION OF THE BROOD CHAMBER

Equipment needed:

- [ ] Hive tool
- [ ] Smoker or hive bomb
- [ ] Frame grip
- [ ] Empty deep super

Step-by-step directions:

- ■ **Dress properly.**
- ■ **Light the smoker** or have hive bomb handy.
- ■ **Enter the hive.** You may not want to smoke the hive entrance.
- ■ **Inspect the brood chamber.** After reading the balance of this section, follow the steps in Unit 12.

This inspection should be brief. As soon as you find eggs in the cells—one egg per cell—you know your queen is at work without having to see her personally. Occasionally a new, young, mated queen will lay two eggs in some cells. Look for the pattern of eggs in almost all the cells in the area in which eggs are seen. Laying workers lay eggs in no particular pattern.

If you find no eggs on this first inspection, you should order a new mated queen immediately. When the new queen arrives, follow the directions for introducing the queen given earlier in this unit. Allow four days for introducing the queen, check for release of the queen, wait seven days, and perform this first inspection of the brood chamber again. (Continue to medicate the hive to control AFB/EFB and Nosema Disease as instructed in Unit 11.)

You should check the hive once a week for several weeks until it is well started. Remember the *Rule of Thumb* for adding another super: When about seven frames are partially drawn and partially filled, it is time to add the next brood super and, later, the next honey super.

Also remember that it is always better to add a super too early than to wait too long, until the super becomes crowded and full. It is usually better to add one super at a time as needed than to add several supers all at one time. Remember that circumstances can alter conditions, and if you are going to be away and cannot tend the hive for some weeks or during the honey flow, add several supers before you go.

Begin the normal routine of brood chamber inspections. Inspect as often as you like, but always check the brood chamber for queenrightness at the following times.

1. When you add your second honey super.
2. After you have removed the honey supers for extracting in the fall, or when you begin your medicating in the fall.

Both of these inspections are discussed in the units in which the manipulations are covered.

UNIT **15**

Swarms and Swarm Control Techniques

WHY BEES SWARM

Swarming is the reproductive cycle in the life of the hive or colony (not the queen). Every species must reproduce, or it will become extinct. Bees in the hives of beekeepers or in the hollows of trees swarm to increase their numbers and thereby increase the chances that the species will survive. Once you realize that the reproductive urge in any species is the strongest urge in nature, you will understand that swarming cannot be controlled easily or entirely, nor would you want it to be. Some swarm control techniques can increase your chances of controlling swarms from your hive. These methods are discussed later in this unit. (See Figure 15-1.)

A DISCUSSION OF SWARMING

Because swarming and swarms will be a continuing problem as long as you keep bees, a somewhat lengthy discussion is needed of what causes swarms, how swarms act, what occurs in the hive that a swarm has left and what occurs in the new home the swarm has found. Following the discussion of swarming is a summary of many of the facts known about swarms and swarming that can be used as a guide in the control of swarms from your hives.

Swarm season occurs each year from several weeks before the honey flow really gets started to several weeks into the honey flow. The swarm season in any area usually lasts from four to six weeks. The time of the swarm season is part of nature's plan; what better time could there be for a new colony (swarm) to get established than when there is a honey flow and lots of pollen for the new unit to build up on and prepare for the first winter?

Once swarming begins, the process is almost irreversible. Swarming is a planned activity; the hive will not swarm unless queen cells are left

Figure 15-1, Some Swarms Ready to Be Captured. Most swarms are quite easy to capture. Each of these swarms was knocked into a cardboard box and then dumped into a proper hive.

so the hive can have a new queen after the swarm (about 40 to 50 percent of the bees in the hive) has left the hive. From five to fifty queen cells in all stages of development are left when the hive swarms. The swarm leaves the hive about the time the first queen cells are capped, unless it is held back by rain or unseasonably cold weather. The old queen cannot fly when she is heavy with eggs, so about three or four days before the swarm issues from the hive, the old queen is put on a diet, which somehow trims her weight rather quickly. Probably she is not fed royal jelly, new eggs are not produced in the ovaries, and the eggs in stock are laid. Slim and trim at the time the swarm issues, she is able to fly out with the swarm. Because she has not flown since mating, at least a year ago in most cases, she tires easily. This no doubt accounts for the fact that the swarm usually settles for the first time within about 100 feet to 100 yards from the hive.

Several days before the hive swarms—but not always—about one or two quarts of bees may be seen hanging about on the front of the hive (some say in a horseshoe pattern). The weather generally is mild at the time swarming occurs, so if you see the bees clustered on the front of the hive during the cooler days of spring or early summer, you can assume that the hive is getting ready to swarm.

Just before the swarm issues from the hive, the bees gorge themselves with honey/nectar. It is a mystery how the bees know who goes with the swarm and who stays with the old hive. Often you will find bees in the swarm with pollen on their legs, and quite often there are many drones with the swarm. Perhaps many drones indicate that the swarm has issued with a virgin queen; that is, that the swarm is a second swarm from the hive.

Gorging permits the bees to take along the food stores and building materials they need for their new home. When a bee's stomach is full of nectar/honey, the wax glands begin to secrete wax flakes, which the bee turns into new honeycomb necessary to begin brood rearing and provide honey storage in the new home.

Let's follow the swarm. (We'll come back to the original hive later.) Swarms issue from the hives between 9:30 A.M. and 3:00 P.M. Most swarms seem to issue between 11:00 A.M. and 1:00 P.M. As the swarm issues from the hive, it tends to drift with the breeze. The air is filled with bees flying in all directions in wide and narrow circles. The bees are gorged with honey, and some bees are carrying pollen, which they must have brought in just before they joined the swarm as it issued. Remember that the queen has not flown in months and will tire easily and land on some object nearby, usually within 100 feet. The queen may leave the hive early in the swarming or near the end of the swarming. When she tires and lands, usually within twenty minutes, the thousands of bees that were in the air will soon cluster on the same object where the queen has landed, usually a branch. The swarm may hang from one-half hour to as long as three days.

While the swarm is hanging, the scout bees leave the cluster and search the area within several miles, looking for a suitable home. As the scout bees return, they report their findings by the communications dance. Somehow the bees decide on one of the locations, probably by the excitement of the dance, which causes other scout bees to check out the location. They in turn dance excitedly about the location. So a choice is made.

All of a sudden, usually at midday, the cluster breaks, the air fills with bees, and the swarm takes off in the direction of the new home. If the new home is some distance away and the queen tires, she will settle again on some object; within about twenty minutes all the bees will settle down, and the swarm cluster will appear. Usually when the queen stops to rest and the swarm clusters around her, they settle in for the night and do not break cluster and move on to the new home until the next day.

When the swarm reaches the new home, the air again fills with bees, but this time they begin to move into their new home. The new home may be an overturned box, a trunk with a hole in it, an old chest with a door ajar, an oversized hole in a stucco wall around a water pipe or electrical wire, or a crack or knothole in the wall of a house where the bees can get inside to the studding, just to name a few.

Watching bees move into bait hives set up in the backyard has shown when and how the swarm gets ready to move in. First only a few bees check out the hive, then some hours later there may be thirty or forty bees, and later there may be many more. When the bees have chosen a new home, a number of bees tend to stay at the location, preventing another swarm from moving in while the main swarm is deciding on the location or is on the way. Finally, on the second or third day (usually in the afternoon), in comes the swarm. The bees fill the air and begin to move into the bait hive.

Within a few hours, most of the bees have moved into the new home and will cluster at the highest point. As the cluster hangs, each bee full of honey/nectar, the bees' wax glands begin to secrete wax flakes, which are lifted off with a spine on the middle leg and brought to the mouth. There the flakes are manipulated and molded into the honey comb with which we are familiar.

As the comb is built, the queen begins to lay eggs and the bees begin to store the honey/nectar that they brought along. In about twenty-four hours, depending on the size of the swarm, the swarm may have drawn over a square foot of comb. Swarms on foundation in a hive often can draw about one frame a day for several days. Older field bees, rid of their load of nectar, begin to search the neighborhood for pollen and nectar, which they gather and bring back to the new home. The new hive begins to prosper as more comb is drawn, more brood is reared, and food stores (nectar and pollen) are gathered for the coming winter.

Back at the original hive. (We'll return to the new home later.) After the swarm leaves, taking approximately half of the field bees and half of the young (house) bees about twelve days old and older, the old hive waits for a new queen to emerge from the many queen cells left. The hive usually must wait from five to seven days after the swarm leaves for the oldest queen cell to release its queen. The first queen to emerge searches out other older queen cells about to release their queens, cuts a hole in each cell, and stings the occupant. Only if two queens emerge simultaneously will the queens meet in mortal combat. After the new queen disposes of rival queens most nearly ready to emerge, the bees begin to tear down the remaining queen cells and dispose of their occupants.

The new virgin queen spends several days maturing and gathering strength. She makes several short orientation flights, then after several days to a week she goes off on her mating flight or flights.

The drones from many hives in the area apparently gather together in the air in a specific locality to wait for the scent of a virgin queen in flight. As the drones on station tire and return to the hive, other drones arrive to maintain the vigil. When the drones detect the scent of a virgin queen on her mating flight, they head in her direction at top speed, and the queen dashes away from their pursuit. In this manner only the strongest and swiftest drones will catch and mate with the virgin queen.

The virgin queen may mate with a number of drones on a single mating flight, or she may make several flights, mating with a few drones each time. When the swiftest drone catches the queen, he grasps her with his legs and contines to fly while carrying her during copulation. Because the drone is a specialist—a flying genital organ—when the genitals are everted at the time of copulation he virtually blows up, driving the penis into the queen's body and forcing the sperm into a receptacle within her body, called the *spermatheca*, where it is stored for future use. After exploding, the drone tumbles backward off the queen and is probably dead before his body hits the ground. In succession, other drones mate with the queen, exploding and adding their sperm to that already in the spermatheca in the queen's body.

The queen must mate within the first few weeks of her life, or she will never mate and will become a drone layer. Because an unmated queen (or a poorly mated queen who runs out of sperm) has no sperm in her spermatheca, she cannot lay fertilized eggs, which become worker bees, or, when needed, queens. An unfertilized egg becomes a drone; the process of birth from an unfertilized egg is known as *parthenogenesis*, a process not uncommon in many insect species.

After the queen mates, several days elapse while the ovaries develop and begin to produce eggs. The queen's diet is now almost entirely royal jelly, because for some time she will be producing almost her own weight in eggs laid each day. The queen never leaves the hive again until the hive swarms but performs her specialist's duty, laying the thousands of eggs needed to maintain the hive population.

It will be five to six weeks after a hive swarms before the first new bees begin to emerge from the brood cells. In the meantime, the honey flow continues (like time, it waits for no man or bee). If the hive is a wild hive in the hollow of a tree, sufficient honey is gathered for the hive to sustain itself for the winter, because the bees will not have to share their honey with humans. However, in the beekeeper's hive, with only half the bees remaining after the swarm and many of those bees dying before the new bees begin to emerge, the honey stores for the beekeeper are drastically reduced. A hive that swarms and is left to produce its own new queen usually will produce only about one-third the honey that the hive would have produced had it not swarmed.

Meanwhile, back at the new home, now a hive. When the swarm leaves the old hive, it takes with it the old queen. It is possible that every swarm, several weeks to several months after becoming established with an old queen, succeeds that old queen in favor of a new young queen. The replacement of an old queen by a new queen without swarming is called *queen succession.* Succession of the queen occurs when the bees in the hive rear a new queen to replace an old or failing queen. The new virgin queen and the old queen are kept apart while the new virgin queen disposes of her other queen-cell rivals and until the new virgin queen has mated and begun to lay. The two queens, the young and the old, may lay together for several days; then the old queen disappears. It is very possible that queen succession can and does occur in the hives without the beekeeper being aware that it is happening.

WHY THE WORLD IS NOT OVERPOPULATED BY HONEYBEES

The world is not overpopulated by honeybees because many things happen that keep every swarm or every hive that swarms from surviving. When a hive swarms, taking the old queen along, the old hive is vulnerable, for the new queen may get lost on her mating flight and not return. There are no stand-in virgin queens waiting in the wings to take over (they are all dead), and there are no larvae left young enough to turn into replacement queens. The virgin queen may be eaten by a bird, fall into a spider's web, get caught in a rain shower, or fly too far from the hive on the mating flight and get lost. She may return to the apiary and go into the wrong hive, where almost certainly she will be killed by the bees in that hive. If for any reason the queen is *lost*, the hive is doomed. The beekeeper can save the hive by requeening with a new mated queen or providing a frame of eggs and larvae so a new queen can be produced in hopes that she will mate and return to the hive.

The swarm itself is vulnerable on the way to the new home, because it may be killed by burning or insecticides, or it may be assaulted and battered by stones and other missiles in the hands of youngsters. If the swarm is accompanied by a virgin queen, it is as vulnerable to the loss of a queen as any other hive having a virgin queen.

In northern areas winters are sometimes so severe that hundreds or thousands of hives may winter kill through no fault of the bees or the beekeeper. In all areas of the country, unexpected flooding may carry away hives. Lack of pollen or nectar sources caused by a late or protracted, cold spring may cause many hives to starve to death just before nectar is available. Indiscriminate use of pesticides by homeowners, farmers, orchardists, and spraying firms can kill thousands of hives each year. Fires destroy many wild hives in forested areas and leave fewer homes for the bees to occupy. Probably in most years, though, more swarms find homes than are killed or die out.

Bees have a tremendous reproductive capacity, and humans can increase the number of hives quite rapidly, if desired, by capturing swarms, dividing hives, producing package bees, and rearing queens. It does not appear that the honeybee will become an endangered species unless and until the human race becomes an endangered species.

MULTIPLE SWARMS FROM THE SAME HIVE

Under normal weather conditions, most hives that swarm will "throw off" only one swarm. However, if the weather is inclement at the time the swarm normally would have issued, with rain or unseasonably cold weather lasting for several days or a week, the hive may throw off two, three, four, or more swarms, called *afterswarms*.

Weather permitting, the prime or first swarm usually issues a day or two after the first queen cells are capped over. Because queens require only sixteen days from egg to emergence, there are only seven days to emergence from the time the queen cell is capped. Those who observe queens very carefully say that at the time the queen leaves the hive with the swarm, she emits a sound, called a *pip*. If the swarm leaves the hive when it normally would, there are no queens in the cells old enough to answer the call of the departing queen. If inclement weather holds the swarm back from three to five days, several queens in the cells near enough to emergence can answer the call of the departing queen. It is suspected that, for example, if three queens in the cells answer the departing queen, then three afterswarms will be thrown off, because the mechanism somehow is set in motion for the queens to be kept apart until the multiple swarms have taken place. A fourth queen will become the mother of the hive. Multiple swarms deplete the hive population considerably, because each swarm takes a portion of the bees remaining after the original swarm took about half of the bees.

QUEEN SUBSTANCE

Queen substance is a glandular secretion given off at the mouth parts of the queen. Queen substance is believed or known to have these effects on the hive.

1. Young queens apparently secrete larger quantities of queen substance than do older queens. Because the queen substance is passed around the hive, it is suspected that as the queen gets older and secretes less queen substance, this triggers the swarm urge in the spring or leads to the succession of the queen at other times of the year.
2. It is known that the hive can tell very quickly when the queen is missing from the hive. Possibly the queen substance is no longer available, and the bees begin to build queen cells around a number of day-old larvae to replace the missing queen. This fact is important to remember when de-queening a hive to install a new mated queen. Do not give the bees any time after de-queening before installing the new queen, or they will start new queen cells and reject (kill) the new mated queen.
3. The queen substance passed around the hive to all the worker bees (incomplete females) inhibits the development of their ovaries and therefore their egg-laying capability. It is known that a hive that has been queenless for several weeks will become a "laying worker hive." This hive is hopelessly queenless, and one or many worker bees who have not tasted queen substance (or if it has worn off) have ovary development and begin to lay eggs. All the eggs laid by laying workers will be infertile and will produce only drones.
4. Queen substance is the scent that attracts the drones to pursue and mate with the virgin queen. It appears that queen substance is the scent that holds the swarm together around the queen. It is known that when the queen is unable to accompany the swarm from the hive, the swarm will return to the hive after about thirty minutes.

SOME FACTS ABOUT SWARMS AND SWARMING

1. The hive will not swarm unless it leaves queen cells so the hive can have a new queen to continue on. The new virgin queen may be lost, and the hive may die out, but not because the bees swarmed away without leaving a potential queen. Do not mistake a hive that absconds for a hive that swarms. Often an absconding hive will cluster and be mistaken for a swarm (which it is as it hangs in the cluster), but if you look in the hive that absconded you will find almost no bees, for they have left honey stores, brood, and everything. If we place bees in an intolerable situation or the situation becomes intolerable, the only recourse the bees have is to abscond to find a more pleasant home environment.
2. The swarm will not leave the area or remain clustered for long if there is no queen in the swarm. This is true whether an old queen or a virgin queen would have gone with the swarm. Do not believe that because the swarm leaves the hive (after you have trapped the queen or used a clipped wing queen) the swarm urge is satisfied when the swarm returns to the hive. The bees will try again and again. If you have a clipped wing queen, after about two tries they will kill that queen and go with one of the virgin queens.
3. The hive will not swarm if all the field bees are removed from the hive. The swarming will be put off until a new population of field bees is present and the same conditions of crowding occur. The swarm is made up primarily of field bees and young bees twelve days old or older.
4. Crowding in the hive is probably the primary cause of most swarming, followed next by an old queen in the hive. To eliminate crowding in the brood area, many experienced beekeepers reverse the brood chambers once or twice in the spring. Others use a third brood chamber temporarily until the honey flow begins, then confine the brood rearing back into two brood chambers. Adding honey supers over a queen excluder does not seem to solve the problem, because newly emerged bees and young house bees do not readily leave the brood area in which they emerged. This is understandable, because that is where their work is to be performed.
5. The older the queen gets, the more certain you can be that the hive will swarm during the next swarm season, if the queen is not replaced by succession in the meantime. Probably the only way you can control swarming with any degree of certainty is to re-queen each year with a new young mated queen, and make sure that the brood chambers do not become crowded.

6. Begin your swarm control methods well in advance of the time the bees begin to make plans for swarming. If the queen is young, and the brood chamber is not crowded as the population is building up, the bees will have no cause to swarm and probably will not swarm. Once the bees have made plans for swarming, it is almost too late for most swarm control methods to work, because the swarm urge, when initiated, is almost irreversible; to control it then is very work-intensive.

   The author finds little pleasure in hunting for and tearing out queen cells, only to find, after doing all that work, that he missed a cell and the bees have swarmed anyway. It is more fun, if they are going to swarm anyway, just to let them swarm.

7. One fact we know about swarms is that the bees are going to swarm, if not every year, then every other year. You cannot keep bees without getting stung occasionally, nor can you keep bees without the hive swarming annually or biennially. Be philosophical about swarms and expect them. Do not worry if you have a few swarms, or if you lose a few. In Unit 17 you will find how to utilize your swarms and other swarms in the neighborhood.

SOME SWARM CONTROL METHODS THAT MIGHT WORK

No swarm control method will work all the time, and most will not work most of the time. Some methods should or could be used together, as is noted in the first method listed. Step-by-step directions for performing the manipulations for these methods of swarm control are detailed after this section.

1. Re-queen each fall or spring with a new mated young queen.

   If you live in the North, you can re-queen either in the spring or fall. If you live in the South, you usually cannot get queens early enough to make re-queening easy in the spring, because the hive is very strong when queens are available. You may find it best to re-queen in the fall, just before, or immediately after, the honey flow ends.

   A new queen will lay heavier and longer into the fall than an older queen might, producing more young bees for the winter clustering. The queen will be young enough not to be a cause of swarming during the following honey season. By re-queening and moving frames of brood above the queen excluder into the first honey super early in the spring (to eliminate crowding), you stand the best chance of having a hive that will swarm only occasionally. If you reverse the brood chambers early (after medicating) and then after a few weeks begin to move frames of brood above the queen excluder, your chances of having no swarms will be even better. The young queen is the most important step, then work the other swarm control methods around the new young queen, whether you re-queen in the spring or fall.

2. Divide the hives.

As long as you want to increase the number of hives in your apiary, you can reduce swarming effectively by dividing your hives and re-queening with new young queens. In the North, young mated queens are readily available by the time you make your divide. In the South, young mated queens may not be available by the time you should make your divide (about mid-March), and you will have to let one of each divide make its own queen. Do not divide hives that have a tendency toward meanness or overaggressiveness. Re-queen mean hives before making divides, or make divides and be sure to re-queen both those divides when queens are available.

In the North: Set the upper brood chamber on another bottom board, and split the brood and stores so the divide that will be in a new location will have an extra frame of brood. Remove the old queen when you find her as you are splitting the brood and stores. Re-queen each divide with a new mated queen. The hive with the least brood will be left on the original site and will pick up the returning field bees, making the population about equal in each hive. If you re-queened your hive last fall, you will not have to re-queen the divide that contains that young queen. In two to three weeks, add a second brood chamber to each divide, then add honey supers as required during the honey flow.

In the South: Set the upper brood chamber on another bottom board, and split the brood and stores so the queenless divide will have half of the brood and will set on the original site. Be sure there are eggs and larvae in the queenless divide, for they will make a new queen. Move the divide with the queen to a new location in the apiary. The field bees will return to the queenless hive to increase the population. This hive will be making a new queen and the first new bee is about forty-five days away, if everything goes right. In the other hive, the queen will continue to lay and build up the population. In two to three weeks, add a second brood chamber to the queenright divide. The queenless divide will have to be checked in three or four weeks to ascertain that the new queen has mated and is laying. If no eggs or larvae are found by the fourth week, order a new mated queen for that divide.

Somewhere down the line you are going to have all the hives you want or can afford, and you will have to look at the other methods

of swarm control. Restocking hives that have died out over winter with frames of brood from your stronger hives or moving frames of brood above the queen excluder into the first honey super might work well.
3. Make a man-made swarm.

This might work well for the beekeeper who has one, two, or three hives and wants no more, does not want swarms in the neighbors' trees, and does not want to lose honey production because of swarming. The man-made swarm is almost identical to the divide mentioned in the previous method (under "In the South"), except that the two divides are reunited, using the newspaper method, just as the honey flow begins. In this way the two units are brought back together, the population of the united hives will be strong, and the hive will produce a full crop of honey during the honey flow. This method permits the beekeeper to re-queen without de-queening, if the queenless portion of the divide raises a new queen. When the two divides are reunited, the new young queen almost always kills the old queen, so the hive will have a new young queen.
4. Remove and place frames of brood above the queen excluder.

Shake frames of brood free of bees (so as not to move the queen with the brood), and place the frames in the honey super above the queen excluder. Replace these frames with empty frames from the honey super. Each week or ten days shake another pair of brood frames and move them above the queen excluder into the honey super. Eventually the frames of brood first moved up will be free of brood and can be replaced in the brood chamber when another pair of brood frames is moved up.

When you move the brood above the excluder, the nurse bees move up and out of the brood chamber, alleviating the crowding, and the queen has new space to fill in the brood chamber. As the brood emerges from the brood combs, the bees will eventually begin to fill them with honey during the honey flow. The only minor disadvantage of this method is that you will have a honey super with four dark or brood combs full of honey when it comes time to extract. This is no real disadvantage, because dark comb holds honey as well as white comb. If you mark the honey super having dark comb when you store it, and use it as the first honey super every spring, you will have only one super with four dark combs for the storage of honey each year. Using this method and re-queening every year is probably the best way for the hobbyist beekeeper to reduce swarming.
5. Reverse the brood chambers.

Some beekeepers practice this method as a swarm control manipulation. Logically it *should* work, but bees are not logical creatures. In a two brood chamber hive, brood rearing usually begins in the upper brood chamber. About six weeks before the honey flow begins, reverse the brood chambers. Presumably the upper brood chamber is full of brood in all stages. When the chambers are reversed, the empty cells not being used in the lower brood chamber are now on top, and the queen will move up and begin to fill those frames. Wait three weeks (one metamorphic cycle of twenty-one days), then reverse the brood chambers again. All the brood should have emerged from the lower brood chamber, which is now on top for the queen to come up and begin to fill those combs. Repeat the reversal in three more weeks, and you should be into the honey flow. With luck, the hive will not swarm if you have been adding honey storage space in anticipation of the honey flow. If your queen is an old queen, this method and most other methods will not stop the hive from swarming.

6. Switch locations of strong hives and weak hives.

If you have a number of hives, some that are weak and others that are strong and may swarm, you might consider changing the locations of a weak hive and a strong hive. What will happen is that all the field bees will return to the old location, where the weak hive now sits. The strong hive has been depleted of all the field bees and should not swarm. The weak hive is strengthened by the field bees from the strong hive. Exchanging frames of brood between the hives might serve the same purpose. After the exchange, begin moving frames of brood above the queen excluder to alleviate crowding in the brood chamber.

7. Tear out queen cells.

If you plan to use this method, you should begin checking the hive at weekly intervals as the swarm season approaches. The easiest way to spot the swarm queen cells is to smoke between the two brood chambers, slide the upper super back (toward you) about one inch, then tip the upper super up about 90°. Many of the swarm queen cells are built along the bottoms of the brood frames and hang down beyond the frame bottom bars. (See Figure 15-2.) Anytime you spot queen cells in these positions, you can assume that the hive is preparing to swarm; you should go into the brood chambers and, frame by frame, search for and tear out all—repeat, *all* —queen cells. If you miss a single queen cell, the hive will swarm. Ten days after you have torn out the queen cells, you must go into the brood chambers again and tear out queen cells. One week later you had better go into the brood chambers and tear out queen cell cups that are again being built for queen rearing.

All this is very work-intensive. It makes lifting out two brood frames at ten-day intervals and moving them above the queen excluder seem like fun.

8. Re-queen with a clipped wing queen.

Some beekeepers re-queen with clipped wing queens, thinking that if the queen cannot fly away, the hive will not swarm. Do not believe this. If you have a queen with a clipped wing, watch the ground in front of your hive every day during the swarm season for a ball of bees about the size of a walnut. When you find this ball of bees, it will tell you that the hive tried to swarm that day. Because the queen cannot fly, the swarm has returned to the hive. Brush the ball of bees onto a piece of cardboard, shingle, or a dustpan, and dump the bees and queen on the hive entrance. On the morrow, remove all the frames in the brood chambers and tear out queen cells. If you do nothing but put the queen back in the hive, the bees will attempt to swarm again, and you will find the ball of bees again in a day or two. If you persist in replacing the queen and doing nothing else, the bees will kill the clipped wing queen and swarm away with the first virgin queen.

Before leaving this discussion, once again remember that you are working against the strongest urge in the universe, the reproductive urge to maintain species survival. The wise beekeeper observes, reads, listens, and *thinks* about behavior characteristics of the bee that can be used to get the bees to do what he or she wants them to do most of the time. There are times when we can use even the most undesirable traits of the bees to our advantage. Think about this as you read through this book and observe your bees.

STEP-BY-STEP DIRECTIONS FOR SOME SWARM CONTROL METHODS

Re-Queening Each Fall or Spring

See Unit 4 for directions on ordering queens. See Unit 13 for de-queening and re-queening the hive.

Making the Man-made Swarm or Making a Divide

Begin the following procedures about six weeks before the honey flow starts. In the South, this will be about mid-March; in the North, this will be about early to mid-June. Both the man-made swarm and the hive divide follow the same steps up to a point. The man-made swarm is united with its other half; the divides are left to become separate hives.

If you want no more than the one or two hives that you have now and don't want swarms to frighten the neighbors, making a man-made swarm each spring should solve most of your swarm problems.

If you want to continue to increase the number of hives in your apiary, you can divide each strong hive each year to double the number of hives and probably halve the honey production per hive.

Equipment needed:

- ☐ Hive tool
- ☐ Smoker or hive bomb
- ☐ Frame grip
- ☐ Queen excluder
- ☐ Empty super
- ☐ Extra bottom and top for each divide or man-made swarm

Step-by-step directions:

- ■ **Dress properly.**
- ■ **Light the smoker** or have hive bomb handy.
- ■ **Enter the hive.**
- ☐ Smoke between the upper and lower brood chambers. Wait one minute. Set the upper brood chamber on the inverted hive top.
- ☐ Shake the bees from each frame in the upper brood chamber onto the frame tops of the lower brood chamber. Set each shaken frame into the empty super. The frames are shaken to get the queen down into the lower brood chamber. (See Unit 13 for steps on shaking bees from the frames.) Brush the bees from inside the upper brood chamber, and set it back on the inverted hive top.
- ☐ Shake the bees from six frames from either side of the lower brood chamber, and set them in the empty upper brood chamber. Place any brood in the center of the brood super.
- ☐ From the empty super, remove and place two frames of capped brood and one frame containing eggs and larvae back into the lower brood chamber. Fill the other three frame spaces with nearly empty frames and/or honey. Save some honey for the other half of the divide.
- ☐ Place the queen excluder on the lower brood chamber, and set the upper brood chamber on the queen excluder. Replace the frames remaining in the empty super into the upper brood chamber, keeping the brood near the center of the brood chamber and the nearly empty frames and/or honey on the outer edges. Close the hive.
- ☐ *Wait twenty-four hours.*
- ☐ So far you have shaken the queen down into the lower brood chamber. The queen excluder over the lower brood chamber keeps the

queen from moving into the upper brood chamber so you know where she is. The nurse bees can and will move up through the queen excluder to tend the brood. In twenty-four hours you can move the upper brood chamber aside onto a hive bottom, with a good supply of bees, knowing the queen is not in that brood super.

The Man-made Swarm After Twenty-four Hours

- ■ **Dress properly.**
- ■ **Light the smoker** or have hive bomb handy. Smoke between the upper and lower brood chambers. Wait one minute.
- ☐ Set the extra hive bottom on the ground nearby. Set the upper brood chamber, with top, on that bottom board.
- ☐ Remove the queen excluder and place the extra top on the lower brood chamber. Set this divide one foot away from the position of the original site with the hive entrance reversed 180°. Store the queen excluder until needed.
- ☐ Set the upper brood chamber hive divide on the original site with no change in hive entrance direction.

What will happen now is this: With the queenless hive on the original site, all the field bees will join the queenless hive, and their population will keep this divide strong while a new queen is being developed. After a few days, the divide with the queen will have a new crew of field bees, the queen will continue to lay, and the divide will continue to grow.

The Hive Divide After Twenty-four Hours

In the North: You should have your new mated queens on hand.

- ■ **Dress properly.**
- ■ **Light the smoker** or have hive bomb handy. Smoke between the upper and lower brood chambers. Wait one minute.
- ☐ Set an extra hive bottom on the ground nearby. Set the upper brood chamber, with top, on that bottom board. Re-queen this divide. (See Unit 13.)
- ☐ Remove the queen excluder from the lower brood chamber. If you did not re-queen last fall, re-queen this divide. Place a hive top on this divide and move it to any other location in the apiary. This is the divide that has the queen, and it will have to be de-queened before re-queening. (See Unit 13.)
- ☐ Move the upper brood chamber divide onto the original site. The field bees will return to this unit and increase its population. Actually, it makes no difference which of the divides is moved to another location in the apiary if you have re-queened both divides.

Check the divides after four days to release the queens. Check the hives one week later to be sure the queen is laying. Add a second brood chamber when required, and add honey supers as required during the honey flow. (See Unit 12.)

In the South: Use same directions as for "In the North" if you can obtain mated queens early enough, but it is doubtful that you can. Otherwise, use this method:

- **Dress properly.**
- **Light the smoker** or have hive bomb handy. Smoke between the upper and lower brood chambers. Wait one minute.
- ☐ Set an extra hive bottom on the ground nearby. Set the upper brood chamber, with top, on that bottom board.
- ☐ Remove the queen excluder from the lower brood chamber and place an extra top on the divide. Move this lower brood chamber divide to any other location in the apiary. Be sure to move this lower brood chamber divide, because it has the queen, and you want the field bees to come back to the queenless divide to bolster the population while the new queen is being developed.
- ☐ Set the upper brood chamber divide onto the original site. Check the queenless hive after four weeks to make sure the new queen has mated and begun to lay. Add a second brood chamber to the queenright hive after two weeks. You also may want to raise frames of brood above the queen excluder.

Weekly Inspection of the Man-made Swarm Divides

Week 1. Queenright hive: Check that the queen is laying.
Queenless hive: Check that queen cells are being developed.

Week 2. Queenright hive: Add queen excluder and first honey super. Raise two frames of brood above the queen excluder into the honey super. See the section entitled "Moving Frames of Brood above the Queen Excluder," which follows in this unit.
Queenless hive: Check that queen cells have developed and that several queens may have emerged.

Week 3. Queenright hive: Raise two more frames of brood above the queen excluder into the honey super.
Queenless hive: Check carefully for eggs and very young larvae in the brood frames. If your queen is laying, you can go to the next section on uniting the two halves of the man-made swarm; otherwise, wait one more week.

Week 4. Queenright hive: Raise two more frames of brood above the queen excluder, unite the two halves of the man-made swarm, and add the second honey super.

Queenless hive: This hive should now have a mated and laying queen. In any event, you will be uniting the two halves of the man-made swarm. If the queenless divide now has a queen, you will have the new young queen as the mother of the united hive. If the queenless hive did not re-queen itself, you will want to re-queen the united hive in the fall, if not sooner.

Uniting the Man-made Swarm Divides

It is now time to unite the two man-made swarm divides that are sitting side by side. It is hoped that each has a laying queen. Proceed with the following steps.

Equipment needed:

- [ ] Hive tool
- [ ] Smoker or hive bomb
- [ ] Single sheet of newspaper
- [ ] Roll of masking tape

Step-by-step directions:

- ■ **Dress properly.**
- ■ **Light the smoker** or have hive bomb handy.
- ■ **Enter the hives** that are sitting side by side. The following steps may seem complicated, but the hive with the old queen must now be turned around so the hive entrance faces in the direction that it faced before you made the divide. You also should put the hive with the new queen on top of the hive that has the old queen. The new queen has a better chance of taking over from this position. She could be killed by the bees from a stronger hive placed above passing through her divide after the divides are united.
- [ ] Enter the hive that has made the new queen. Place the hive top on the ground nearby. Set the brood super on the inverted hive top.
- [ ] Enter the hive with the old queen. Place the inverted hive top on the ground nearby. Smoke and remove the honey super and queen excluder from this hive. If you have not moved two frames of brood above into the honey super in the last week, do so now.
- [ ] Place the bottom board left free against the hive with the old queen. Set the brood super with the old queen over onto that bottom board. Now the hive will have the entrance facing in the same direction it did before the hive was divided some weeks ago, and you are ready to unite the brood chamber with the new queen on top of the brood chamber containing the old queen.

- [ ] Place a single sheet of newspaper over the brood chamber with the old queen, the one now resting on a bottom board. If there is much wind, tape two sides of the newspaper down so it will not blow away while you are lifting the other brood chamber up to set it on the paper. Punch about six holes in the paper with a hive tool, pencil, or nail.
- [ ] Set the brood chamber with the new queen on top of the newspaper. Place the queen excluder on this second brood chamber, and set the honey super (containing the brood frames, if you were moving frames of brood into the honey super) on the queen excluder. Close the hive. You may find it necessary to add a second honey super, depending on how much honey has been stored in the first honey super. Store the extra top and bottom.
- [ ] Proceed from this point with the normal inspections of the honey supers and manipulation of frames in the supers.
- [ ] You may notice quite a bit of confusion around the hive for about two days. The reason for this is that the hive entrance has been changed for the one unit, and those field bees are looking for the entrance in the rear. They will find and relearn the new entrance, so do not worry about the confusion.

Moving Frames of Brood above the Queen Excluder

Equipment needed:

- [ ] Hive tool
- [ ] Smoker or hive bomb
- [ ] Frame grip
- [ ] Bee brush
- [ ] Queen excluder
- [ ] Deep honey super with frames of comb or foundation
- [ ] Empty super

Step-by-step directions:

- ■ **Dress properly.**
- ■ **Light the smoker** or have hive bomb handy.
- ■ **Enter the hive.** Place the inverted hive top on the ground nearby.
- ■ **Inspect the upper brood chamber,** which is probably where the brood will be in a two brood chamber hive. If the brood is not in the upper brood chamber, set the upper brood chamber aside on the inverted hive top and inspect the lower brood chamber.
- [ ] Set the honey super with frames of comb or foundation on the inverted

hive top or on a pair of two-by-fours. Remove the two center frames from the honey super, and set them in the empty super or stand them on end nearby.
- [ ] Remove two frames of brood, preferably frames with about half capped brood and half eggs and larvae. Shake the bees off over the brood chamber to be sure the queen is not moved above the queen excluder. Place the two frames of brood removed and shaken in the center of the honey super.
- [ ] Push the remaining brood frames together, and place the empty frames from the honey super on one side of the brood chamber in frame spaces 3 and 4 (or 7 and 8). Do not split up the brood frames in the brood chamber.
- [ ] Place the queen excluder on the upper brood chamber (after setting it on the lower brood chamber) if it was removed to get to the brood in the lower brood chamber. Set the honey super containing the two frames of brood on the queen excluder.
- [ ] Close the hive. The nurse bees will hurry through the queen excluder to tend the brood in the two brood frames raised up into the honey super.
- [ ] *Wait ten days to two weeks.*
- [ ] Repeat the previous steps, raising two more frames of brood above the queen excluder and replacing the brood frames moved up with frames of comb or foundation from the honey super.
- [ ] *Wait another ten days to two weeks.*
- [ ] Repeat the previous steps, raising two more frames of brood above the queen excluder. This time remove the first two frames of brood comb raised up into the honey super. Most of the brood in those two frames will have emerged by this time, and the frames can be returned to the brood chamber from which they came.
- [ ] It may be necessary to add a second honey super at this time, so have one ready, just in case.

Note: If you mark this first honey super in which you have raised the brood frames and use this super for the first honey super every year, you will only have one super per hive with four dark combs. In other words, use the dark combs over and over, year after year.

Switching Locations of Strong Hives and Weak Hives

Equipment needed:

- [ ] Hive tool
- [ ] Smoker or hive bomb

- [ ] Hive staples, four per level
- [ ] Hammer
- [ ] Screen roll to close hive entrance
- [ ] A friend, hand truck, wagon, etc., to move the hives

Step-by-step directions:

■ **Dress properly.**
- [ ] At dark or just after dark, slip a roll of screen into the hive entrance of all the hives to be exchanged. If you use nylon or plastic screen, you can cut the screen roll with scissors to the proper length at the time you put the screen in the entrance. To get most of the bees that are hanging on the front of the hive or out on the entrance inside, use a puff or two of smoke and wait five minutes.
- [ ] After the screen rolls are in place, hammer the staples in place on each hive to be moved. Staple the bottom board to the lower brood chamber and the upper brood chamber to the lower brood chamber. (See Figure 9-3 for the placement of the staples at each level.) It is not likely that there will be any honey supers on the hive at this time; however, if there are or if you move the hive for some reason when a honey super is on the hive, it should be stapled down also.
- [ ] Pick up and move the weak hive by carrying it, moving it with a hand truck, etc., and place it beside the strong hive it is to be exchanged for. Pick up and move the strong hive, and place it on the location from which the weak hive was moved. Return and put the weak hive on the exact location from which the strong hive was moved.
- [ ] When all hives are exchanged, go back and pull out the screens. You may want to drop them in front of the hive as you pull them and pick them up the next time you visit the apiary. The staples can be left in the supers until you have a need to remove them, then use the hive tool to lift out the staples. Try not to use a flashlight at night while working with the bees, for some bees may fly out and fly to the light, land on you, and crawl up inside your clothing.
- [ ] What will happen during the next day is this: Field bees from the strong hive will be returning to the original location on which the weak hive now sits. They will know something is wrong, but after some confusion, they will join with the weaker hive, adding their population to that hive. The same thing will occur at the site of the weak hive, but fewer field bees will be returning, because the hive was weaker. What really is important with this move is that it causes the strong hive, which might be getting crowded and will soon prepare to swarm, to lose all its field bees, and the hive will not swarm until it builds up its field strength again. You would probably only use this method if you got behind in swarm control

some year. If you use this method, begin to move frames of brood above the queen excluder to keep the hive from becoming crowded again too soon and swarming.

Reversing Brood Chambers

Equipment needed:

- [ ] Hive tool
- [ ] Smoker or hive bomb
- [ ] Pair of two-by-fours

Step-by-step directions:

- ■ **Dress properly.**
- ■ **Light the smoker** or have hive bomb handy.
- ■ **Enter the hive.** Smoke between the upper and lower brood chambers. Wait one minute.
- [ ] Place the inverted hive top and the two-by-fours on the ground nearby. Place the upper brood chamber on the inverted hive top. Place the lower brood chamber on the pair of two-by-fours. Clean up the bottom board with the hive tool and set it back in place.
- [ ] Place the upper brood chamber on the bottom board.
- [ ] Place the lower brood chamber on the upper brood chamber now on the bottom board. The brood chambers have now been reversed. Close the hive. Add honey supers as required during the following manipulations.
- [ ] *Wait three weeks.*
- [ ] Follow the previous steps for the second reversal of the brood chambers. If you have added a honey super, modify the fourth step as follows:

    Set the honey super off on the inverted hive top, set the queen excluder on the honey super, then set the upper brood chamber on the queen excluder. Place the lower brood chamber on the pair of two-by-fours. Return to the fifth step in this section.
- [ ] *Wait three more weeks.*
- [ ] Follow the previous steps for the third reversal of the brood chambers. If you have not added a honey super by this time, you should add one now. Close the hive and continue to add honey supers as required during the honey flow.

Tearing Out Queen Cells

This is the most work-intensive and least effective of all swarm control methods, and yet it is the one most often mentioned for the control of swarming. You must continue to check for queen cells between the brood chambers of the hive as the swarm season approaches, then tear out the queen cells. Repeat the process at ten-day intervals because the bees will make new queen cells once you have torn out the first set. If you miss a single queen cell at any time in this process, you will have a swarm, and all your work will have gone for naught.

Finding the Queen Cells

As the swarm season approaches, you must check the hive regularly for queen cells. Most swarm queen cells are found near the bottom bars in the super in which brood rearing is taking place. If your brood rearing is in the upper brood chamber, you can smoke between the brood supers, slide the upper brood super toward you (work from behind) about one inch, then tilt the brood super up and hold it vertically while you look at the bottom bars. When you see queen cells hanging, you must then go into the hive, remove all frames, and tear out or destroy the queen cells in both brood chambers. If your brood is in the lower brood chamber, set the upper brood chamber aside, tilt the lower brood chamber up for a look, then, if you see queen cells, go into the brood chambers and tear them out. (See Figure 15-2.)

If you find no visible queen cells, make the same inspection at weekly intervals, setting any honey supers aside if you have added them. (Remember that you can move the bees on the frame by gently blowing on them.)

Steps for Tearing Out Queen Cells

Equipment needed:

- ☐ Hive tool
- ☐ Smoker or hive bomb
- ☐ Frame grip
- ☐ Pen knife
- ☐ Pair of two-by-fours
- ☐ Empty super

Step-by-step directions:

- ■ **Dress properly.**
- ■ **Light the smoker** or have hive bomb handy.
- ■ **Enter the hive.** Place the inverted hive top on the ground nearby.
  It is presumed that you have found queen cells in the hive and are now ready to tear them out.
- ■ **Inspect the brood chambers.** (See Unit 13 for the steps.) You will be looking for queen cells as you inspect the brood frames. Before

SWARMS AND SWARM CONTROL TECHNIQUES 141

Figure 15-2, Queen Cells. This pair of queen cells, located near the bottom of the frame, extend down between the bottom bars. In hives that are ready to swarm or that have swarmed, a number of queen cells will be hanging as shown here. By tipping the brood chamber up to near vertical, the queen cells are easy to spot. Note: All queen cells hang vertically wherever they are located on the brood frames, whereas the workers and drones are reared in the near-horizontal brood cells.

you start looking for queen cells, you will set the upper brood chamber on the inverted hive top. Any honey supers can be set aside on the pair of two-by-fours.

☐ Set the upper brood chamber on the inverted hive top.

☐ Begin with the lower brood chamber. Remove one frame at a time. Shake the frame over the lower brood super to remove as many bees as possible so they will not be in the way as you search for the queen cells and queen cell cups. Be sure to look carefully in the corners where the comb is cut away, for queen cells often hang in those areas and are easily missed. If you want to move the bees from an area, blow on them gently and they will begin to disperse.

☐ With the tip of a hive tool, pen knife, or fingernail, tear out or cut into each queen cell that you find. If you want to look into any open queen cells or queen cell cups, turn the frame over so you can look down into them. If you want to taste royal jelly, now is a good time because there will be copious amounts in each queen cell that is not capped. After destroying the queen cells on each frame, set the frame in the empty super nearby. Set the inner cover over the super to retain heat for the brood. When all ten frames have been checked over carefully and all queen cells removed, replace the frames in the lower brood chamber.

- [ ] Repeat these steps for the upper brood chamber, shaking the bees off the frames into the lower brood chamber. Return all frames to the upper brood chamber and close the hive. If you have not added a queen excluder and a honey super by this time, you should do so to alleviate crowding somewhat. It is not too late to raise frames of brood above the queen excluder into the honey super.
- [ ] *Wait ten days.* Repeat the previous inspection steps to remove any newly developed queen cells. If you find queen cells on this inspection, make the next queen inspection in one week. If your bees swarm between these inspections, you can be sure that you have missed a queen cell in the hive.

Buying a Queen with a Clipped Wing

A queen with a clipped wing will not stop swarming, but she will let you know that the hive has tried to swarm. If you check your hive each day during the swarm season and do not take any other control activity, such as moving frames of brood above the queen excluder, one day you will find a ball of bees on the ground in front of the hive. This tells you that the hive tried to swarm, but your queen could not fly away with the swarm, so the bees all returned.

Finding the queen with a clipped wing outside the hive lets you know that the hive has tried to swarm, so it is time to go into the hive and tear out all the queen cells as directed in the previous method. If you find the queen out in front a few days after tearing out all the queen cells, you know you missed a queen cell or two. If you find the queen out in front in about ten days, you will know the bees have made a second set of queen cells, and you must tear them out again.

Repeatedly returning the queen with a clipped wing to the hive without tearing out the queen cells will result in the killing of the clipped wing queen by the bees, who will then swarm away with one of the virgin queens that has emerged.

A suggestion: Use a clipped wing queen in conjunction with other swarm control methods—then you have a back-up to tell you if the swarm control methods used did not work and you must tear out queen cells.

IF YOUR HIVE SWARMS

Do not be devastated. It is important to remember to inspect the brood chamber of a hive known to have swarmed three or four weeks after the swarm left to insure the virgin queen has mated, returned, and is laying. Many hives do not re-queen after a swarm issues. A hive without a mated queen becomes, first, a laying worker hive and, second, a wax moth-infested hive, unless the beekeeper discovers the queenlessness early enough to take corrective action. If you find no eggs or larvae in any brood frames four weeks after a hive has swarmed, order a new mated queen for the hive immediately.

UNIT **16**

Capturing and Hiving Swarms

When a swarm issues from a hive, the bees are gorged with nectar/honey. Once settled in the swarm with the queen, they have no home to protect until they find a new home. In the meantime, the swarm is the most docile of any unit of bees. Capturing a swarm of bees is a relatively safe and simple task that can be accomplished by almost anyone.

More often than not, swarms settle in the yards of nonbeekeeping homeowners, who know only that bees sting and make honey, and who fear the sting more than they enjoy the honey. The swarm is an awesome sight, with tens of thousands of bees in the air all at one time before they settle on a branch or shrub. Most beekeepers try to help the communities in which they live by removing swarms in their neighborhoods. It is generally a free community service. (See Figure 16-1.)

GETTING ON A LIST FOR SWARMS

Some of the places you can leave your name for swarms might be:

- Police or sheriff's department
- Fire department
- Health department
- County communications (In some counties this group serves the police, fire, and sheriff's departments.)
- Beekeepers in the area
- Beekeeping associations
- Exterminators (Many do not like to destroy bees.)

If you want only one or two swarms, be ready to *get off the list* when you have them, or you will be getting dozens of calls. Do not give beekeepers a bad name by taking swarm calls and doing nothing about

CAPTURING AND HIVING SWARMS 145

Figure 16-1, Capturing Swarms from Trees. Get the box up under and around the swarm if possible, give the branch several sharp raps, and you have captured a swarm. Wait about twenty minutes for most of the bees to settle in the box with the queen, then tape it shut and punch nail holes in the side for ventilation. (See Figure 16-2.)

them. If you take the time to get on a swarm list, take the time to get off when you have all the swarms you want.

BEFORE YOU GO OUT TO CAPTURE A SWARM

Get the address and phone number of the person who has knowledge of the swarm before you go out to capture it. Call first to verify that the swarm is still there, even if it has been only an hour since the bees settled. Some swarms leave before you reach them even if they were there when you left home. Do not waste your time going for a swarm when all you know is that it is on such and such a street, between here and there. If you cannot pin it down with an address, forget it.

When you call to get more information about the swarm, ask these questions:

1. How high is the swarm and what is it on? If the swarm is above fifteen feet or in a dangerous position, do not try to get the swarm. Tell the people the swarm will leave within a few days. In most cases, people would rather wait than pay your hospital bills. Do not risk an accident for a swarm of bees. There will be many more swarms in easily accessible locations.
2. Does the caller have a ladder long enough to let you reach the swarm? If not, you will have to carry a ladder—always a nuisance.
3. How long has the swarm been there? This might give you an idea of when the swarm may move.
4. How many other people have been called to pick up the swarm? You may be one of dozens called.
5. Has the swarm been sprayed with insecticide or washed down with water? Forget those.
6. Can you cut a branch to get the swarm, if necessary?
7. Is the swarm in or on a building? Ninety-five percent of swarms on buildings are either entering the building or belong to hives that have gone unnoticed in the buildings for some time.

Sometimes homeowners do not know or do not tell the whole truth about the bees in the yard, and you may discover on arrival that the bees are, in fact, a hive that has been in the house for some time. The homeowners may have learned from previous calls that most beekeepers do not bother with bees in a structure, so when you ask they insist that it is a swarm.

Many homeowners cannot tell five feet from twenty feet. "Of course you can reach the swarm," they say, but when you get there you find that they must have thought you were twenty-five feet tall.

If you find that the swarm is in a dangerous position, do not try to take it. Ask the homeowner to call someone properly equipped or to wait a few days for the swarm to leave. Do not attempt to take bees out of a house, hollow tree, sign, etc. Most beekeepers are foolish enough to try it once, but we smarten up pretty quickly. The bees are just not worth what it costs in time, gas, stings, and general inconvenience. If you need bees that badly, you will find that it is quicker and cheaper to buy a package of bees. As a rule, there are so many swarms around that it is silly to try to get a hive out of a permanent location. Unless you can bring it home, realy want to do it, or will get paid a lot, forget about those. If you must try it, see Unit 28.

There are swarms that have located themselves in boxes, barrels, birdhouses, barbecues, chests, trunks, etc. If the hive can be closed up, lifted into a truck, and brought home (all the above after dark), then it is worth having, for you can take the bees and comb out at your convenience, after the bees have oriented to your yard. See Unit 29 for removing bees from these odd hives.

CAPTURING A SWARM

Equipment needed:

- [ ] Coveralls
- [ ] Gloves
- [ ] Bee brush
- [ ] Veil
- [ ] Boot bands
- [ ] Branch clippers
- [ ] Cardboard box, about 12" × 18" × 18", with flaps that close completely
- [ ] Roll of masking tape, 1" or 2" wide
- [ ] 20 feet of light rope
- [ ] 8-penny nail

Capturing a Swarm on a Supple Branch

Step-by-step directions:

■ **Dress properly.**
- [ ] With the flaps of the box open, slip the box around the swarm. Get the box up around the swarm as far as possible. Give the branch a sharp rap with the heel of your hand or your fist. Most of the swarm will fall into the box; then give the branch several more raps to knock all the bees loose.

- [ ] Turn the box on end. Hold the box or set it on a ladder, stool, etc., about two feet below the branch where the swarm was hanging. The many bees that flew into the air will begin to enter the box and recluster. If you are holding the box, hold it for about two minutes, then slowly lower it to the ground. Set the box on end with the flaps open. If you have the queen in the box, the bees will stay. On the flaps of the box, you will see a number of bees with their tails high, heads down, and wings fanning. This is a good sign, for it means the queen is in the box and the bees are scenting to call the other bees to join them. (See Figure 32-1.)
- [ ] Fasten a light rope around the branch on which the swarm was hanging, and give the rope a sharp jerk about every minute to get the 10 percent that go back to the branch to answer the scent call at the box.
- [ ] Wait about twenty minutes after knocking the swarm into the box. Most of the bees will have joined the cluster in the box. Brush any bees on the outside of the box over the flaps and into the box. Close the flaps and tape the flaps closed. With the eight-penny nail, punch about forty holes in each side of the box for ventilation. Brush any bees on the outside of the box away, and put the box in the front or back seat of your car or in the trunk. (See Figure 16-2.)
- [ ] There may be many bees left that will cluster back on the branch, but you cannot get them all unless you stay until dark. Tell the homeowners that the bees will leave or die within a day or two. Tell them that if they are concerned or worried, they can spray those left with insecticide at dark.
- [ ] Before leaving with the swarm, check that you have your masking tape, bee brush, and everything else you brought with you.
- [ ] If you stop to pick up another swarm before getting this swarm home, it would be a good idea to set this swarm out of the car in the shade while you capture the next swarm.
- [ ] When you get the boxed swarm home, place it in a cool place—in the garage, perhaps. Place the box with the taped side down, and the bees will cluster on the upper side of the box. This will make dumping the bees into the hive easier because they will not be clustered on the flaps of the box when they are opened. Try not to hold the swarm in the box for more than forty-eight hours.

Capturing a Swarm on the Ground

■ **Dress properly.**
- [ ] Place the box on its side near the swarm, and push a flap several inches under the edge of the swarm. With the bee brush or a twig,

CAPTURING AND HIVING SWARMS 149

Figure 16-2, A Captured, Boxed Swarm Ready to Bring Home, Store for a Few Hours, or Dump. If you carry and store the box with the tape side down, the bees will hang from the highest point and will be easier to dump, because they will not be hanging from the flaps as you open the box. Notice the nail holes in the box sides to provide ventilation.

flick a few bees into the box, and the swarm will begin to run into the box.

☐ Wait about twenty minutes. Set the box upright, then brush the bees on the outside of the box over the flaps into the box. Close the flaps and tape them closed. Punch about forty holes in each side of the box for ventilation. Brush aside any bees on the outside of the box, and place the boxed swarm in your car.

☐ Check that you have everything you brought along. When you get the swarm home, set it in a cool place, taped side down, until ready to hive the swarm. Try to hive the swarm within forty-eight hours.

Capturing a Swarm on a Solid Object

Follow these steps to capture swarms from a post, fire plug, traffic light, tree trunk, etc.

■ **Dress properly.**

☐ Look for the largest (thickest) mass of bees on the object. Hold a flap of the box under the bees and brush them onto the flap and into

the box. Move the box around the object, brushing bees onto the flap and into the box.
- [ ] Set the box with the flaps open on its side about one foot from the bottom of the object. Brush all the bees off the object onto the ground or into the air. The bees on the ground will run into the box, and those bees in the air will answer the scent call to the box.
- [ ] Continue to brush any bees that return to the object into the air for the next ten to twenty minutes.
- [ ] After about twenty minutes, turn the box up and brush the bees on the sides of the box over the flaps into the box. Close the flaps and tape the flaps closed. Punch about forty holes in each side of the box with the nail. Brush away any bees remaining on the outside of the box, and place the boxed swarm in your car.
- [ ] Check that you have everything you brought along. When you get the boxed swarm home, set it in a cool place, taped side down, until you are ready to hive the swarm. Try to hive the swarm within forty-eight hours.

Capturing a Swarm on a Large Branch

Follow these steps to capture a swarm on a branch that is too large to rap or shake.

■ **Dress properly.**
- [ ] The bees usually cluster around the branch and may be along several feet of the branch. Often a portion of the swarm will be hanging below the branch.
- [ ] Hold the box under the swarm. With the handle of a bee brush or a dry twig, slice through the swarm along the bottom of the branch. Still holding the box, begin to brush or push the bees into the box from the top and sides of the branch.
- [ ] Continue to hold the box, or set it on a ladder or stool while you brush all the bees from the branch into the air to get them to answer the scent call to the box. Brush vigorously to get all the bees from the branch. Keep the bees from landing on the branch again by brushing the bees into the air during the next ten to twenty minutes. If you have to hold the box, after about five minutes slowly set it on the ground, on end with the flaps open.
- [ ] After twenty minutes, brush any bees on the outside of the box over the flaps into the box. Close the flaps and tape the flaps closed. Punch about forty ventilation holes in each side of the box with the

nail. Place the boxed swarm in your car, after brushing away any bees on the outside of the box.
- [ ] Check that you have everything you brought with you. When you have the boxed swarm home, set it in a cool place, taped side down, until you are ready to hive the swarm. Try to hive the swarm within forty-eight hours.

Capturing a Swarm on a High, Supple Branch (12 to 20 feet)

This method requires a twenty-foot extension ladder, preferably aluminum, and a cardboard box about fifteen inches wide.

■ **Dress properly.**
- [ ] Because the uprights on the extending portion of most ladders are about fourteen inches apart, force the box between the uprights of the ladder down onto the top rung of the ladder. Leave the flaps of the box free and open.
- [ ] Stand the ladder up under the swarm, and raise the extending portion until the box is under and around the swarm.
- [ ] Lift the ladder and shove it quickly against the branch once or twice. With a good, smart shove, the swarm should be knocked from the branch into the box.
- [ ] Lower the extending portion of the ladder about two rungs, and hold the ladder upright under the branch on which the swarm was hanging so the bees in the air and on the branch will be able to answer the scent call to the box. Hold the ladder in this upright position for about ten minutes.
- [ ] Continue to hold the ladder upright, but lower the ladder one rung at a time at one-minute intervals. Moving the box down slowly will allow more of the bees to follow it down.
- [ ] When the extending portion of the ladder is all the way down, walk the ladder down to the ground, close the flaps on the box, and tape the flaps closed. Pull the box out from between the uprights of the ladder, and punch about forty holes in each side of the box. Brush away the bees on the outside of the box, and place the boxed swarm in your car.
- [ ] Check that you have everything you brought with you. When you get the boxed swarm home, set it in a cool place, taped side down, until you are ready to hive the swarm. Try to hive the swarm within forty-eight hours.

HIVING A SWARM

When the swarm is captured and brought home, the sooner it is hived, the better. Swarms can be hived at any time of the day, but they seem to settle in more quickly if they are hived about one hour before dark. With darkness coming, the bees settle in for the night. One reason you want to get the swarm into the hive as quickly as possible is that the swarm begins to draw comb from the top of the box. It is better to get the comb drawn on the foundation than from the top of the swarm box. Sometimes you have no alternative but to hold a swarm for more than forty-eight hours; if this becomes absolutely necessary, the following section will describe how to do it.

Holding a Swarm in a Capture Box for More than Forty-eight Hours

☐ If you know that you must hold a swarm in the capture box for more than forty-eight hours for some reason, place the boxed swarm in the exact location where your hive will set when you hive the swarm.

☐ With a sharp pen knife or utility knife, cut a one-inch high by three-inch long opening in the box about one inch up from the bottom on one end. Have the taped side of the box down. Here is an easy way to cut the opening without letting too many bees out as you do it: Cut the sides and top of the opening, then tape over them with masking tape. Now make the lower cut of the opening and pull off the tape, bringing away the cutout from the opening.

☐ When the box has an opening and is in the place where the hive will set, the bees orient, eliminate, and ventilate. When you are ready to hive the swarm, most of the field bees will have oriented to the location and will merely be coming back into the hive rather than into the box. The only disadvantage here is that the bees will begin to draw wild comb from the top of the box, and if the swarm flies from the box for very long, eggs will soon be laid in the wild comb. But if you have some good reason for not hiving the swarm within forty-eight hours, you can expect to pay for it by having to waste the comb that should be drawn on foundation but that now hangs from the top of the box. By feeding sugar solution to the swarm after hiving it you can keep them drawing comb at a rather rapid pace. If you wait too long holding the swarm, refer to Unit 29 for directions on how to transfer the bees to a proper hive and salvage the brood in the wild comb.

☐ When you are ready to hive the boxed swarm that you have been holding, simply move the box aside and place the hive exactly where the box has been sitting, dump the bees from the box into the hive (see the directions that follow), close the hive, and continue to feed sugar solution until the swarm is well settled in.

Top Dumping to Hive a Swarm (See Figure 16-3.)

Equipment needed:

- [ ] Boxed swarm of bees
- [ ] Feeder and quart jar of sugar solution **or** feeder and one gallon of Nosema-medicated sugar solution
- [ ] Scrap block of wood
- [ ] Empty super

Step-by-step directions:

- ■ **Dress properly.** You should not need a smoker, unless you have held the swarm, as mentioned above, for more than forty-eight hours.
- [ ] With your hive in its permanent location, remove the top, inner cover, and three middle frames. Stand the frames nearby, or set them in an empty super.
- [ ] Bring the boxed swarm from the cool place where it has been stored. Be careful not to disturb or dislodge the cluster in the box.
- [ ] Set the box on the hive while you remove the tape holding the flaps closed. Remove all the tape so that the flaps will fall free when you are ready to dump the bees into the hive.
- [ ] While holding the flaps closed, lift the box about one foot above the hive and give the box a quick upward jerk. Most of the cluster will fall into the hive among the frames. Knock or whack the sides of the box several times as you hold it over the hive to remove the last of the bees in the box. Toss the box about ten feet away. Later, remove the box from the area, for the scent may attract many of the bees back to the box.
- [ ] Gently replace the three frames removed earlier. Set them gently on the bees, and the bees will move out from under them. *Do not under any circumstances leave out a frame when hiving the swarm.* If you do, that open area is where the swarm will cluster and begin to draw comb, attaching it to the inner cover or hive top.
- [ ] Bees will be all over in the air by this time, but do not worry; they will all return to the hive by dark.
- [ ] Replace the inner cover and hive top.
- [ ] Insert the feeder base in the hive entrance at one side. Invert a quart jar of sugar solution into the feeder base. Keep the feeder jar filled until at least five pounds of sugar has been used, or mix

154 KEEPING BEES

Figure 16-3, Top Dumping a Swarm. Remove several frames and dump the swarm on and between those remaining. Replace all frames that were removed and set the inner cover on the hive. Leave the top off until after dark; the flying bees will return to join the swarm through the front and the oval hole in the inner cover.

the Nosema-medicated sugar solution and feed that. (See Unit 11 for preparing the Nosema-medicated sugar solution.) Close the hive entrance to about three inches alongside the feeder with a scrap block of wood.

Do not disturb the hive for seven days.

☐ If you worry that the swarm may abscond or leave the hive during the next day or two—and they sometimes do—make a roll of screen that fits the hive entrance between the feeder and the opposite side of the bottom board. If you use nylon or plastic screening, you can cut the roll to size with scissors at the time you push the roll in. Insert the screen in the hive entrance after dark. Leave it in until one hour before dark the next evening, and pull it out. The bees will fly out to orient and eliminate and will be back in the hive by dark. If you are a real worrier, insert the screen again after dark, and leave it in until about one hour before dark the next evening. This helps because the swarm will be held in, and all the time they will be drawing comb and storing away the nectar/sugar solution so they should figure that this is going to be home.

Skip ahead to the section entitled "Inspection after Seven Days."

Front Dumping to Hive a Swarm (See Figure 16-4.)

If you want to see your queen as she runs into the hive or capture the queen from a swarm (to replace her with a new mated queen or just to get her out of a swarm to join the swarm with another hive), you can front dump the swarm and let the bees run into the hive through the hive entrance.

Equipment needed:

- [ ] Boxed swarm
- [ ] Bee brush
- [ ] Feeder and quart jar of sugar solution **or** feeder and one gallon of Nosema-medicated sugar solution
- [ ] Scrap block of wood
- [ ] Materials to make a ramp to the hive entrance

Step-by-step directions:

■ **Dress properly.**
- [ ] With the hive in its permanent location, space the frames evenly in the hive, and replace the inner cover and the hive top.
- [ ] Make a ramp up to the hive entrance with plywood, cardboard, or a sheet, etc. Be sure the ramp leads the bees into the entrance. If the queen goes under the hive rather than in it, all the bees will cluster below with the queen, and you will have to rebox and redump the swarm.
- [ ] Gently bring the boxed swarm from the cool place where you have stored it. Do not disturb or dislodge the cluster in the box.
- [ ] Place the boxed swarm on top of the hive, and gently remove the tape from the flaps so that the flaps will fall free when the bees are dumped.

156 KEEPING BEES

Figure 16-4, Front Dumping a Swarm. A large towel has been laid out so the bees do not have to move through the grass but over it. Above: The bees have been dumped about three feet in front of the hive, and a few bees have been tossed onto the hive entrance, where they begin to scent. (See Figure 32-1.) Center: The bees are running into and onto the hive. Below: Notice the gloved finger pointing to the queen as she runs to the hive.

- [ ] Hold the flaps closed and move to the ramp, about two or three feet in front of the hive entrance. Give the box a jerk (if the flaps are down), or invert the box and jerk or dump (if the flaps are up). Knock or whack the sides of the box several times to get all the remaining bees out of the box. Toss the box about ten feet away.
- [ ] Flick some bees up the ramp with a bee brush or twig, and the bees will begin to run up the ramp into the hive. Watch the bees as they run up the ramp, and in most cases, you will spot the queen moving among the bees to the entrance. As more and more bees reach the hive entrance, traffic slows down and the bees begin to move onto the front and sides of the hive. Do not worry about this, for they will slowly drift into the hive overnight.
- [ ] Many bees will be in the air at this time, but do not worry; they will be back in the hive by dark. Place the feeder base in the hive entrance on one side, and close the entrance to about three inches alongside the feeder with a scrap block of wood. Keep the feeder jar filled until you have used at least five pounds of sugar in a sugar solution or feed the Nosema-medicated sugar solution. (See Unit 11 for directions on preparing the Nosema-medicated sugar solution.) *Do not disturb the hive for seven days.*

Inspection After Seven Days

After seven days of uninterrupted activity in the hive, it is time to check the brood chamber to see if the queen has begun to lay. If the queen with the swarm was the old queen, you should find a fair amount of eggs and young larvae in several brood frames. If the queen was a virgin queen from a secondary swarm or a swarm that has killed and left the hive of a clipped wing queen, then you probably will not find any eggs or larvae on this first inspection. Check again in one week. If you do not find eggs or young larvae at that time, order a new mated queen, because your queen was most likely a virgin and was lost on her mating flight.

Equipment needed:

- [ ] Hive tool
- [ ] Smoker or hive bomb
- [ ] Frame grip
- [ ] Empty super
- [ ] Terramycin®/powdered sugar mix
- [ ] One gallon Nosema-medicated sugar solution
- [ ] Pollen supplement patty

Step-by-step directions:
- ■ **Dress properly.**
- ■ **Light the smoker** or have hive bomb handy.
- ■ **Enter the hive.** Place the inverted hive top on the ground nearby, and set the empty super on it.
- ■ **Inspect the brood chamber.** (See Unit 12.) This inspection should be brief. As soon as you find eggs in the cells—*one egg in each cell*—you know your queen is at work without having to see her.

    If you find no eggs, make this same inspection in one week. If you still find no eggs on the second inspection, order a queen for immediate delivery. When the new mated queen arrives, re-queen as instructed in Unit 13.
- ☐ Begin to medicate the hive to control AFB/EFB, as directed in Unit 11. If needed in your area, place a pollen patty on the frame tops and spoon the AFB/EFB-medicated mix on the frame tops around the pollen patty. If you have not begun to feed the Nosema-medicated solution, begin to feed it when you next fill the feeder jar. (See Unit 11.)
- ☐ Close the hive. Check the brood chamber weekly for the first few weeks. Remember the rule of thumb that says when about seven frames are partially drawn and partially filled, it is time to add the next brood chamber or honey super. It is better to add more space too early than too late, particularly when the honey flow is coming soon or has already started. *Remember:* Do not medicate when honey supers are on the hive. Stop medicating about ten days before you add any honey supers.
- ☐ Begin the normal routine of brood chamber inspections. (See Unit 12.) Inspect as often as you like but at least as often as suggested in Unit 12.

Capturing the Queen When You Front Dump a Swarm

Remember that you cannot leave the queen out of the hived swarm for hours or days. Get the queen in a cage and back in the hive within half an hour. If the queen is not returned to the hive in a timely fashion, the bees may abscond or join the nearest queenright hive. The queen catcher makes it possible for you to catch the queen, tie a string to the queen catcher, and push it back into the hive entrance. The queen is not away from her swarm for more than one minute. The queen catcher is worth every cent it costs at a time like this. (See Figure 16-5.)

You can make a queen cage of screen, or use a queen cage that you received a queen in earlier. To make a queen cage, wrap a three-inch

Figure 16-5, The Queen Catcher. The queen catcher open and closed. The bees can move through the slots of the queen catcher as they do through the queen excluder.

square piece of screen around a half-inch diameter piece of dowel, and fold in one end. Pull out the dowel and sew the screen together with a strand of the screen or fine wire, such as framing wire or fine copper wire. Herd the queen in the open end, then fold it over, and you have a quick homemade queen cage.

If you want to capture the queen from a swarm you front dump so you can replace her with a mated queen (which you are going to order by tomorrow), follow these steps.

Equipment needed (in addition to that needed to hive a swarm):

☐ Queen catcher, previously used queen cage, homemade queen cage, or small jar

Step-by-step directions:

☐ As the bees run up the ramp after the swarm has been front dumped, watch for and capture the queen in a small jar or queen catcher.
☐ If you have a queen catcher, go to the next step.

Carry the jar with the queen into the bathroom so you can herd her into the previously used queen cage or the homemade cage. Turn on the light over the medicine cabinet. Set the queen loose on the medicine chest mirror, then herd her very gently into the cage. Try not to pick up the queen, for you might damage her.

When the queen is in the homemade cage, fasten the ends of the cage securely so the queen cannot get out. Then fasten two feet of fine wire or string to one end of the homemade cage.

When the queen is in the previously used queen cage, make several turns around the edges of the cage with cellophane tape to close off the exit holes on the ends. Place a thumbtack on the end of the cage, and fasten two feet of fine wire or string to the thumbtack.

☐ Return to the hive, and push the queen cage into the entrance of

160 KEEPING BEES

the hive and back about the length of the hive tool. Be sure the screen side is up if you are using a wooden queen cage. If you have a queen catcher, fasten two feet of string or fine wire to the handles and push the queen catcher back into the hive entrance about ten inches. Leave the string or wire dangling out of the hive entrance.
☐ When your new queen arrives, de-queen the hive by pulling on the string or fine wire, drawing out the queen in the cage or queen catcher. You will then be ready to re-queen the hive with no trouble.

Remember: You cannot use the queen catcher to add a strange queen to a swarm or hive, because the bees will kill a strange queen unless she is introduced so the bees will be kept from getting to her for about four days. If you want to introduce a queen into a hive or swarm, other than the one she came from, gently herd her into a previously used queen cage or a homemade cage.

UNIT 17

What to Do with Swarms

Because bees will swarm occasionally no matter what, you should turn this undesirable behavior trait to your advantage. You have some idea of how to control swarms (Unit 15), but if the control methods do not work, you know how to capture the swarms (Unit 16). The question now is what to do with a captured swarm.

They say you cannot have your cake and eat it too, but with swarms you can come pretty close. Most beekeeping books say very little about utilizing swarms, but bees do swarm—if not from your hive, from hives in structures, in hollows of trees, or from the hives of other beekeepers. How can you utilize swarms, other than using a swarm to start a new hive? The following ideas might be of some value to the beekeeper who has access to swarms. Each of these ideas is discussed in more detail near the end of this unit, after several sections that will familiarize you with methods of uniting swarms.

1. Use several swarms in the same hive to make one strong hive that will produce a maximum crop of honey the first year.
2. Use a swarm to stengthen a weak hive. This adds about 20,000 bees overnight and gets a weak hive going again.
3. Use a swarm to replace the bees that swarmed from one of your hives. You cannot put the swarm from a hive back into the same hive, but you can put it in another hive.
4. Use a swarm to re-queen a queenless hive.
5. Use a swarm to bring a laying worker hive back to production.
6. Use a swarm to clean up a hive that shows the first signs of wax moth infestation.
7. Use a swarm to clean up a dead hive that is moldy and has hundreds of dead bees in the cells.
8. Start an observation hive with a small swarm.

SOME REASONS SWARMS ARE NOT UTILIZED BY MANY BEEKEEPERS

1. A swarm may be diseased.

    A package or a hive of bees that you purchase may also be diseased. You will seldom find a diseased hive that can throw off a four-pound or larger swarm. Small swarms could carry a disease, but not necessarily. In any event, it would seem foolhardy to hive a package or a swarm without taking the precautions to medicate the newly established hive to control foulbrood diseases.

2. A swarm is wild and more aggressive.

    There is no such thing as a domesticated bee or hive of bees. All bees are wild in the sense that they can live in the woods and forests as happily as they can in the expensive hive furnished by the beekeeper. A swarm can go from a lovely hive in the best area of town to a hollow tree in the woods and continue to thrive. About 95 percent of the bees in the U.S. are Italian bees, and the other 5 percent are Caucasian and Carniolan strains. There are almost no wild bees in this country anymore. However, meanness in bees is a dominant trait, and bees left to themselves in the wild over many generations probably would be meaner than those from the queens you buy individually or that come in packages today. The queen in the swarm that you pick up around town or in the country may be the gentle queen that someone bought last year and was lost in a swarm this year. Then again, the queen in the swarm you capture may be hereditarily mean. Over the years, the author has found that most swarms are not overly mean or overly aggressive. You take what you get and make changes with a new queen later if the hive tends to be more aggressive than you like.

3. The swarm has an old queen.

    If the swarm you capture is the prime or first swarm from the hive, then it has an old queen. If the swarm is a secondary, tertiary, etc., swarm from a hive, then the swarm has a young virgin queen who must mate from your hive. Young mated queens of good quality are available and can be obtained for re-queening swarms as well as established hives. See the discussion for capturing the queen from a swarm in Unit 16.

    If crowding becomes critical, any hive can be forced to swarm whether it has a new mated queen or an old queen. However, with an older queen, the hive will tend to swarm more quickly if she is not succeeded by a young queen after the swarm settles into its new home.

4. A swarm is a nuisance to capture and hive.

Since the cardboard box was invented in the early 1900's, capturing a swarm has become a snap! With the invention of masking tape to seal the box, it is even more of a snap. You can get a proper box at most grocers' for free, and for the cost of a little gas you can gather a swarm of bees for free. Early in the swarm season, the average swarm is usually larger than the package of bees, and in some areas, such as in the south, swarms are available much earlier than packages of bees.

5. Package bees are easy to get.

Package bees are easy to get only if you can get them early enough to build up a strong colony. Packages are best in the north where swarms occur after the middle of May. Package bees are not necessarily best in the South, if the swarm season begins by the first of April, when package bees usually are not available. Early swarms usually are quite large, giving you four to six pounds of bees to get the hive off to a good start. The thing to remember is that mated queens are also readily available, and it is possible later to re-queen a swarm-made hive. Sixty days after re-queening with a gentle productive queen, your hive will be populated by the offspring of the new gentle queen. Also remember that a mated queen costs about one-third as much as a package, and queens are easier to get as the season progresses. Any bees can be used to get a hive started. Your mated queen will produce her own kind, and the swarm bees will all be dead within sixty days.

METHODS OF UNITING SWARMS

Dumping a swarm directly into an existing hive almost always results in fighting and the loss of many bees. Uniting swarms, as in uniting hives, is necessary to give the bees a chance to join together gradually.

One method for uniting swarms is the direct dumping of bees in front of the hive at full dark. This method is similar to front dumping a swarm to start a new hive but differs in that the swarm is dumped in front of an existing hive and it must be done at full dark. This method will be called Carrier's Swarm-Uniting Method. It works nearly as well as the newspaper method of uniting, which is the other method for uniting swarms. It is the same method used for uniting hives for any reason (see Unit 24) and can also be used for uniting swarms.

The author has never heard or read about uniting swarms as discussed in Carrier's Method, and it seems strange that this technique of uniting swarms has never been discovered and used before. The author accepts recognition for this swarm-uniting method, by virtue of the copyright of this book, until an earlier copyright for this method is discovered.

Both methods can unite swarms with existing hives without removing the swarm queen, but there are times when the swarm queen should be removed. If you have an expensive queen and do not want to lose her, remove the queen from the swarm. If you want to unite a swarm in a hive after the honey super is on the hive, remove the swarm queen before uniting the swarm in the hive; that is, before dumping the swarm in a honey super over a sheet of newspaper.

Removing the Queen from the Swarm

Before proceeding, give serious consideration to obtaining a queen catcher (see Figure 16-5), or make a queen trap large enough so that you can get the queen in it easily without having to touch and perhaps damage her. The trap, which will have to be made by you, should have a lid that will close securely so the queen cannot get out once you have trapped her inside. The bees must be able to smell and taste or touch the queen. One idea might be a clear plastic pill bottle about one and one-half inches in diameter and two inches deep in which have been drilled many 1/8-inch holes. The queen catcher or your queen trap should have several feet of string attached so it can be removed from the swarm just before the swarm is dumped.

Remember, if you are going to remove the queen from the swarm sometime before dumping the swarm, you must capture the queen in the catcher or queen trap and place her back in the swarm until the swarm is dumped. The bees will become excited or demoralized if they cannot smell, taste, or touch the queen for any length of time.

Equipment needed:
- [ ] Empty cardboard box
- [ ] Boxed swarm
- [ ] Queen catcher or queen trap of some sort
- [ ] Piece of plywood, cardboard, or a sheet

Step-by-step directions:
- ■ **Dress properly.**
- [ ] Set up a box with the flaps open as near as possible to the hive into which the swarm will be dumped. Place plywood, cardboard, or a sheet in front of the box over the lower flap.
- [ ] Remove the tape from the boxed swarm so the flaps will fall free when the swarm is dumped. Dump the swarm about two to three feet in front of the box. Whack the sides of the swarm box smartly

WHAT TO DO WITH SWARMS   165

several times to knock out all the bees, then toss the box about ten feet away.

☐ Flick a few bees toward the open box. As the bees run to the box, watch for the queen moving along to the box. When you spot the queen, get her into the queen catcher or the queen trap cage. Set the queen catcher or closed queen trap cage toward the back of the box with the string dangling out of the box. (See Figure 17-1.)

☐ After most of the bees have run into the box, close the flap to a small opening so the bees in the air can return to the box. Let the bees settle into the box over the next few hours or until you are ready to dump the swarm into the hive.

Carrier's Swarm-Uniting Method

This swarm-uniting method has not been tested exhaustively, but in over 100 unitings it has failed only twice. Several customers in the

Figure 17-1, Dumping a Swarm to Capture the Queen. A large towel has been placed on the ground in front of the cardboard box and over the flap of the box so the bees will run into it. Left: The beekeeper is poised with the queen catcher, watching for the queen. Right: The beekeeper has caught the queen (and many bees) in the queen catcher. A string will be attached to the queen catcher, and it will be placed back in the box until it is ready to be dumped. When the bees are dumped, you can remove the queen from the swarm by lifting the queen catcher out of the box.

shop have been asked to try this method. Some have been amazed at how well it works, whereas others have had it fail on occasion.

You must be aware that, as a rule, two different hives or swarms of bees cannot be mixed together without the bees fighting. Both units of bees feel they are being invaded by robbers, particularly if it is in the middle of the day. There are exceptions to most rules, and it appears that with this method you stand a better than 90 percent chance of uniting a swarm with an existing hive.

The most important thing to remember with this method is that *the swarm should be united at full dark,* not at sundown or dusk. If you are going to use this method, you should check the approximate time that full dark occurs in your area during the swarm season.

Equipment needed:
- [ ] Ramp to entrance of existing hive
- [ ] Smoker or hive bomb
- [ ] Boxed swarm

Step-by-step directions:

Unless you have removed the queen from the swarm, do not attempt to put a swarm back into the hive from which it came. Otherwise, this uniting can be done with the queen in the swarm. Do not use a flashlight or other outside lighting.

- [ ] When darkness approaches, ■ **Dress properly.**
- [ ] Set up the ramp to the hive entrance.
- ■ **Light the smoker** or have hive bomb handy.
- [ ] *Not more than ten minutes before full dark,* smoke the hive entrance with four to six puffs of smoke.
- [ ] Bring out the boxed swarm so as not to disturb or dislodge the cluster. Remove the tape carefully and completely so the flaps will drop free when the swarm is dumped. Bees will escape but do not worry about them. Lift out the caged queen, if you captured her earlier.
- [ ] Lift the box and give it a jerk so the swarm will drop on the ramp between two and three feet in front of the hive. Give the box several smart whacks on the sides to knock out the rest of the bees, and toss the box about ten feet away.
- [ ] Flick a few bees up the ramp, and the bees will begin to run up the ramp, into and onto the hive. As traffic slows down at the entrance, most of the bees will move onto the front and sides of the hive. The bees in the air should return to the hive rather quickly, as it will be almost full dark by this time.
- [ ] The bees on the front and sides of the hive will enter the hive gradually during the night.

Why This Method Works

1. As darkness approaches, the hive does not expect to be invaded so it is not vigilant in anticipation of robber bees. Smoking the entrance will send the guards away to begin to fill up with honey/nectar.
2. As the bees run up the ramp into the hive, the entrance will become clogged, and the mass of bees will gather on the outside of the hive. The bees then enter the hive gradually during the night.
3. The bees in the swarm are gorged with honey. Bees gorged with honey, if challanged by the bees in the hive, will cower and offer up a sip of nectar, which will tend to make them welcome.

You should be aware that this method will not work every time. It may work fifty times before something goes wrong and fighting occurs, or it might happen the first time you try it. Beekeepers do not like to see thousands of dead bees in front of the hive when fighting occurs, but sometimes we can rationalize that "if we did not try to save the swarm, they might all have been killed by the exterminator."

Uniting Swarms Using the Newspaper Method

If you do not like the previous method or you find it does not work, you can use the newspaper method of uniting swarms. Uniting hives using the newspaper method has been done for many years, and it is almost 100 percent sure. Now that you know how to get the queen from a swarm, you can unite a queenless swarm in a honey super above the brood chamber. This is an advantage that Carrier's Method does not permit. *Perform the newspaper uniting about thirty minutes before dark.*

Equipment needed:
- ☐ Boxed swarm
- ☐ Brood super or a super that will be used as a brood super *or* a honey super or a super that will be used as a honey super, complete with frames of comb or foundation
- ☐ Sheet of newspaper and a roll of masking tape
- ☐ Hive tool
- ☐ Smoker or hive bomb

Step-by-step directions:
- ■ **Dress properly.**
- ■ **Light the smoker** or have hive bomb handy.
- ■ **Enter the hive** into which you are going to unite the swarm. Invert the hive top on the ground nearby, and set the empty super with frames of comb or foundation on it. Start this uniting procedure about thirty minutes before dark for best results.

☐ Pick the hive configuration that applies:
   1. The hive has one brood chamber, no honey supers, or one or more honey supers.
☐ You will use a second brood chamber into which you will dump the swarm. The swarm can be queenless or can have the swarm queen—your option.
   2. The hive has two brood chambers, no honey supers.
☐ You will use a queen excluder and a first honey super into which you will dump the swarm. You have no option—the swarm must be queenless when it is dumped, or you will have a queen above the queen excluder laying in the honey supers.
   3. The hive has two brood chambers and one (or more) honey super.
☐ You will use another honey super into which you will dump the swarm. You have no option—the swarm must be queenless when it is dumped, or you will have a laying queen in your honey supers.
☐ Place a single sheet of newspaper over the brood super or honey super, depending on the hive configuration you choose. Tape the edges of the paper to the hive so that it will not blow off while you are setting the super into which the swarm will be dumped onto the newspaper. Punch several slits in the paper with a hive tool, pen knife, pencil point, etc.
☐ Set the brood or honey super on the sheet of newspaper.
☐ Remove three or four frames from the super on the newspaper, and space the frames apart so the swarm can fall between the frames when they are dumped.
☐ Bring out the boxed swarm gently so as not to disturb or dislodge the cluster. Remove the tape so the flaps can fall free when the swarm is dumped *or* pick up the boxed swarm that was made queenless and is placed beside the hive into which it is to be united (the swarm you dumped to capture the queen). Remove the caged queen from this swarm.
☐ Dump the swarm from the box into the super on the hive by giving the box a jerk or, perhaps, first knocking the bees into a corner of the box and then dumping them into the super. Give the box several sharp whacks on the sides, then toss it about ten feet away.
☐ Replace the frames removed earlier and space them evenly in the super. Replace the inner cover and stand the hive top beside the hive.

If you have no inner cover, replace the hive top so there will be space for the flying bees to return to the super in which they were dumped. In case 1, you will dump the bees into a second brood chamber placed on a sheet of newspaper. In case 2, add the newspaper and the queen excluder. Dump the queenless swarm in the honey super and replace the inner cover. In case 3, add the newspaper over the present honey super. Dump the queenless swarm into the honey super added and replace the inner cover. The top is left off so the airborne bees in the swarm will be able to return and join up with the swarm.

☐ After dark, return to the hive and replace the top. The bees in the swarm must work their way through the paper and pass through the hive to fly. While the newspaper is being turned to lint and small pieces, the bees do not fight among themselves. Within a short period of time the two sets of bees will become one. Over the next few days the bees will completely tear up and carry the paper outside the hive; the only trace will be the paper on the outside of the hive.

What about Swarm Queens?

If you unite a swarm with a queen into a hive with a queen using either method, the Carrier or the newspaper, nature will take its course. One of the two queens will be killed by the bees themselves or by the two queens meeting in mortal combat. Only one queen will emerge alive.

HOW TO UTILIZE SWARMS—YOURS AND OTHERS

At the start of this unit, a number of ideas were listed that might be used by the beekeeper who has access to swarms. These are discussed in more detail now that you are familiar with the methods of uniting swarms.

In any of the following directions, when it says dump the swarm onto paper, it is assumed that you will have frames of comb or foundation in the supers involved. You will remove several frames to dump the swarm, but be sure to replace them.

1. *Use several swarms in one hive.* This will make one strong hive that will produce a maximum crop of honey the first year. This idea can be expanded to include starting two or more new hives, each being given up to three swarms per hive as the swarms become available.

Starting multiple hives from swarms works as follows. Suppose you want to start three new hives using swarms. First obtain the hive equipment and have the hives ready in their permanent locations. As swarm season begins, swarm capture 1 goes into hive 1; swarm capture 2 goes into hive 2; swarm capture 3 goes into hive 3; swarm capture 4 goes into hive 1; swarm capture 5 goes into hive 2; swarm capture 6 goes into hive 3; swarm capture 7 goes into hive 1; swarm capture 8 goes into hive 2; and swarm capture 9 goes into hive 3.

Three swarms per hive should be the maximum, unless some of the swarms are very small. You can get too many bees in the hive and force the hive to swarm from crowding. The hive supering will be handled as follows. When you add the second swarm to the hive, add the second brood chamber. When you add the third swarm, add the queen excluder and the first honey super. Use foundation in the supers in preference to comb, because the swarms are primed for comb building (all gorged with honey/nectar).

Watch your honey supers carefully during the honey flow. Remember that you have added about 40,000 bees within a matter of about three weeks, and this hive can draw comb and bring in nectar so fast you won't believe it. Do not get caught short of storage space.

Using Carrier's Method, each swarm (with queen) is dumped in front of the hive as directed, just a few minutes before full dark. Occasionally a swarm that is dumped may swarm back out of the hive the next day. If this happens, or to prevent it from happening, you may want to remove the queens from all but the first swarm dumped into the hive.

Using the newspaper method, the first swarm is dumped into the empty hive. Dump the second swarm, with queen, into the second brood chamber. Dump the third swarm into the honey super above the queen excluder after the queen is removed. In summary, the second swarm is dumped onto paper over the first brood chamber, and the third swarm is dumped onto paper over the queen excluder.

2. *Use a swarm to strengthen a weak hive.* Adding a swarm to a weak hive will increase overnight the population of the hive by many thousands of bees.

Using Carrier's Method, dump the swarm with queen in front of the hive a few minutes before full dark.

Using the newspaper method, add a queenless swarm onto paper above the second brood chamber or onto paper above the present honey super on the hive (after having removed the queen from the swarm).

3. *Use a swarm to replace a swarm.* You can capture the swarm from a hive, remove the queen, and put it back in the same hive. If you have a swarm that gets away, you can capture a neighborhood swarm, remove the queen, and unite it with the hive that swarmed. It would be a good idea to save or bank several of the queens you remove from swarms that you dump. A hive that swarms gets a new queen, but she will not always make it back from her mating flight. It is good to have lots of bees by adding a swarm back into the hive, but it is even more

important to have a laying queen. If you bank several queens and a virgin queen is lost, you can introduce one of your banked queens, who will start to lay a couple of days after the four-day introduction period. (See "Banking Queens" at the end of this unit.)

Using Carrier's Method, dump the queenless swarm in front of the hive. No newspaper is needed to dump the queenless swarm back into the honey super or into the brood super, if the bees came from that hive. If the swarm is from a different hive, use the newspaper method and add the queenless swarm onto paper above the second brood chamber or above the present honey super on the hive.

4. *Use a swarm to re-queen a queenless hive.* If swarms are available, use either uniting method. Otherwise, order a new queen and re-queen as directed in Unit 13.

Using Carrier's Method, dump the swarm with queen in front of the hive a few minutes before full dark.

Using the newspaper method, dump the swarm, with queen, onto paper in the second brood chamber among the bees that are there, or dump the swarm onto paper in a super, with frames, added above the brood chamber. One week later, remove the super added above the brood chamber (unless it was a second brood chamber), shake the frames to remove the queen, then insert the queen excluder and use the super as a honey super.

5. *Use a swarm to bring a laying worker hive back into production.* If swarms are available, use one of these methods. Otherwise, refer to Unit 26 on laying workers.

Using Carrier's Method, dump the swarm with queen in front of the hive a few minutes before full dark.

Using the newspaper method, dump the swarm with queen onto paper in the second brood chamber among the bees that are there.

6. *Use a swarm to clean up and restart a moth-infested hive.* This applies to a hive with just the first signs of wax moth infestation. Badly infested hives should be cleaned up and have the foundations replaced before you add a swarm to the hive. If swarms are not available, see Unit 27. You have no time to lose.

Using Carrier's Method, dump the swarm with queen in front of the hive. It is not suggested that the newspaper method be used. If you do use it, refer to item 5. It is best to have the swarm begin to clean out the wax moth larvae.

7. *Use a swarm to clean up a dead hive.* Always check a dead hive for AFB/EFB.

In the North: Hold the dead hive in a bee-tight area, or close it off so bees cannot rob it out. If the hive is disease-free, add a swarm to the hive and the bees will clean it up till it shines, even if the combs and pollen are moldy and there are hundreds of dead bees in the cells. Before dumping the swarm in the hive, you should inspect for AFB/EFB, brush off the dead bees, and clean up the bottom board. Dump the swarm in front of the hive, or top dump the swarm in the hive as directed in Unit 16.

In the South: If you find a dead hive, you must place each brood super and each honey super in a plastic bag with PDB moth crystals until swarms are available. Air the brood super or supers for several days before dumping a swarm in them. Front dump or top dump the swarm in the hive. Check for AFB/EFB before dumping a swarm in any dead hive.

8. *Start an observation hive with a small swarm.* If you want to start an observation hive—and in the North you will have to restart it each spring—you can set it on the sidewalk, raise or remove the glass from one side, and dump the swarm beside the observation hive about thirty minutes before dark. By dark, most of the bees will be inside. Put the glass back in the observation hive gently, replace and secure the top, brush off any bees still on the outside, and set or hang the observation hive in its permanent location. See Unit 33 for step-by-step directions for starting and manipulating the observation hive.

BANKING QUEENS

Queens can be kept *(banked)* for several weeks by using a modified frame holding up to a dozen queen cages, each containing a queen. This frame, or several frames, is placed in a queenless hive or nucleus. Because the bees have a queen (a frame or more of them), the hive does not start queen cells but tends all the queens. Additional frames of brood—older larvae and covered brood—are occasionally given to the hive from other queenright hives to maintain young bees in the queenless bank hive.

For the hobbyist beekeeper, a number of caged *mated queens* could be kept for several weeks under a rimmed cover on the frame tops of a strong queenless nucleus hive. Add a frame of mostly covered brood about once a week. *Do not* bank virgin queens, because they must mate within the first few weeks or they lose the urge and will become drone-laying queens. Queens may die banked in such a system, usually from chilling in a weak nucleus or from your damaging the queen when you herd her into the cage. It is also possible to bank caged queens above the queen excluder of a queenright hive for a few days.

PART V

Late Spring, Summer, and Early Fall Activities

The next five units cover the activities and manipulations relating to the honey flow--from the beginning of the honey flow to a few weeks after it ends. Activities include adding supers, removing supers, extracting honey, cleaning up extracted supers of comb, and storing extracted supers.

- Unit 18   DURING THE HONEY FLOW
- Unit 19   REMOVING HONEY SUPERS
- Unit 20   PROCESSING AND STORING COMB HONEY

174   KEEPING BEES

- **Unit 21**   EXTRACTING HONEY
- **Unit 22**   CLEANING AND STORING SUPERS OF EXTRACTED COMB

UNIT **18**

During the Honey Flow

Nectar flow or honey flow are phrases coined by beekeepers to describe the period of time when the most blossoms in an area are open and ready for pollination. Pollination is necessary in order for floral species to develop seeds and reproduce so that they do not become extinct.

Pollination occurs in a blossom when the mature pollen from the anther of that flower is transmitted to the stigma of that flower or to a different flower of the same variety. Flowers accomplish pollination in many ways. In some cases, the pollen is produced in a male flower and must be transported to the female flower for pollination to occur.

Most flowering weeds, flowers, shrubs, and trees have colored blossoms to attract the pollinators—insects, primarily; birds and mammals to a lesser degree. Once the pollinators have been attracted, the flower must provide a sweet nectar and/or pollen. Plants that offer the most and the sweetest nectar or the most pollen are the ones that get first call on the pollination service of birds, bees, and other insect pollinators. This period, when most flowering plants are offering nectar and pollen in abundance, is called the nectar or honey flow. The flow period in most areas will last six to ten weeks; then it will dwindle, but the bees will continue searching for even small amounts of nectar or pollen.

Plants that cannot compete with the crowd of plants in the honey flow have adapted themselves in other ways to keep the species alive. Some have become hardy souls that peek up through the last snows of winter to attract the bees as they begin their brood build-up, when even small amounts of nectar and pollen are welcome. Some plant varieties have become late bloomers and are available with small quantities of nectar and pollen after the main flow, and some bloom through the first light frosts of fall.

Most fruit trees are early bloomers and, as a rule, do not produce nectar or pollen in large quantities or of extreme sweetness, but they come into bloom at the time the bees are beginning the intense brood build-up. As a result, they are worked heavily by the bees, because both nectar and pollen are in great demand at that time of the year.

True nuts, grasses (grains and corn), and most nonflowering trees, including the conifers, depend on the wind for their pollination and, as a result, produce very large quantities of pollen (light pollen). The bees collect pollen from many of these pollen-producing plants that produce no nectar.

THE HIVE BEFORE THE HONEY FLOW

In mid-January or early February, limited brood rearing begins in the hive, using the honey and pollen stores within the cluster. As the weather moderates into spring, the brood area begins to expand among the cluster. When the temperature reaches about 60° F and the bees become active for a part of each day, they begin to forage for nectar, pollen, and water. Water becomes quite important in the spring when nectar is not available in the field, because the bees must water the honey down to a nectar consistency to feed it to the larvae. Because brood build-up begins some time before nectar and pollen are available in the field, it is essential that these stores in the hive do not become depleted. At least two brood chambers should be used for winter stores in both the north and the south. Depending on the weather in your area, you may want to carry a third super of honey stores on the hive during the winter in the north. If the hive lightens considerably, take steps to feed the hive. As soon as the hive can be worked, get a pollen supplement patty onto the frame tops. In many cases, the bees make it through the winter only to starve while rearing brood in the few weeks before the early nectar source is available. As more eggs are laid and more honey is fed to the larvae from the brood cells nearby, the queen begins to fill those empty cells with more eggs, more cells are emptied of honey, and more brood is reared.

You should realize that there is such a thing as *chilled brood*. This is caused by (1) the beekeeper keeping the hive open too long when the weather is cold, so the heat in the brood chamber is lost, or (2) natural causes, such as a drop in the outside temperature. If the weather has been pleasant for several weeks and then becomes very cold, the bees will cluster more tightly together. If the bees cannot cover all the brood, the brood outside of the cluster will chill and die. This is usually not disastrous, but if the weather warms up rapidly you will begin to smell the rotting carcasses in the hive and worry that you have a foul-brood disease. Chilled brood may happen several times in a very unseasonable spring.

From the time the fruit trees blossom, more and more nectar and pollen are available in the field, and brood rearing continues at a faster and faster rate. Eventually, the hive may become too crowded and may swarm. In this period, swarm control measures should be taken. Adding a queen excluder and the first honey super will not normally suffice, because young bees do not tend to leave the brood area unless brood is moved. There is nothing to attract the bees to the honey super unless there is nectar to be stored. At this time, all the incoming nectar usually is being fed directly to the larval brood or stored around the brood for the nurse bees to feed overnight.

It is a good idea to have the queen excluder and the first honey super on the hive several weeks before the honey flow is expected to start so they will be there when excess nectar begins to come in. Do not expect that adding that super is going to keep your hive from swarming, because the crowding is going to continue in the brood chambers. Even adding several honey supers would not keep the hive from swarming; although the bees would have plenty of room, none of it would be in the brood area, where it is needed. If you are going to perform any kind of swarm control manipulations with your hive—except tearing out queen cells (it is too late then)—you must begin before the hive becomes crowded and the bees begin *their* swarm planning.

THE HIVE DURING THE HONEY FLOW

If your hive does not swarm and you continue to add honey supers in a timely manner, you will be amazed at how quickly the supers fill. You must check the honey supers at least every ten days during the honey flow until you become familiar with the flow in your area and know when to add supers. Skimping on honey storage space may cause the hive to swarm.

The length of the honey flow depends on your area. The crops grown in the area could cause a longer or shorter honey flow from one year to the next. The heavy part of the honey flow could last about six weeks in northern Maine to more than four months in southern California. It is more than likely that more honey will be stored in the six weeks in Maine than in the four months in California.

In general, the shorter the growing season, the more urgent the pollination requirements of the plants. The more urgent the pollination requirements, the larger the quantity of immensely sweet nectar the plants produce. The more bees you have in the hive, the more quickly the honey supers will be filled.

If your hive swarms just before the honey flow and you do not restock it as directed in Unit 17, your hive will produce only about one-third of the honey it would have produced had it not swarmed. Just before and during the honey flow is the time to get those frames of foundation drawn into comb. When the bees have to hold nectar with no place to store it, they hang around, and the wax glands begin to secrete wax flakes. You will be amazed at how fast foundation can be drawn into comb, filled, and capped.

THE HIVE AFTER THE HONEY FLOW

At some point after the honey flow diminishes, incoming nectar is no longer taken into the honey supers but is stored in the brood supers. As smaller quantities of nectar come into the hive, the bees feed the queen less royal jelly, and she begins to lay fewer and fewer eggs. Soon the death rate of old bees overtakes the emergence rate of new young bees, and the population of the hive begins to decrease. At this time, all the nectar and pollen are being stored in the brood area, leaving the queen smaller and smaller areas in which to lay. The hive eventually will stabilize at about 20,000 to 25,000 bees for the winter cluster. During this time, the drones are driven from the hive, for they will not be needed until spring, when more will be reared. Most of the bees in the winter cluster will be young bees that can live through the winter and bring the hive back into brood activity in the early spring.

Much of the honey stored in the brood area will be uncapped and be more readily available to the cluster during the winter. Winter is one of the most critical periods in the life of the hive. With too many bees and drones, the stores may not last, and the hive will starve.

Do not consider keeping the bees warm when you think of wintering your bees. Bees cluster and generate their own heat inside the cluster. Do very seriously consider venting the hive during the winter. If not vented from the hive, water vapors from condensation of moisture in the air and the metabolic processes of the bees can cause severe stress on the cluster. Bees can stand lots of cold but they cannot stand cold and moist conditions.

In the North: The bees may be clustered for as long as five months. If there are three or four calm, pleasant days during that time that will warm the hive enough for the bees to break cluster and move to new honey stores, the hive probably will make it through the winter. If there are no pleasant days for a period of from six to ten weeks, it is very probable that you will lose the hive to winter kill, which is beyond your control; in the spring you will find your hive dead of starvation while surrounded by honey stores. If the cluster cannot break and move, it will use up all the honey under the cluster and then starve.

PREPARING TO ADD HONEY SUPERS

Rule of Thumb for Adding Supers (Brood or Honey)

(This rule of thumb applies whether you are starting a new hive or supering an established hive.)

- When seven frames of foundation are partially drawn into comb and partially filled, it is time to add another super.
- When seven frames of comb are partially filled, add another super.
- When seven frames in a brood chamber are partially drawn and filled, it is time to add a second brood chamber or, if it is the second brood chamber, the queen excluder and the first honey super.

Once again, remember that a strong hive can draw and fill combs very quickly during the honey flow. The speed with which a super is filled depends on the stage of the honey flow. Generally, the first super is added several weeks before the honey flow begins, or earlier if you are going to move brood up as a swarm control method. This first honey super will not fill very rapidly. The second and third supers usually will fill rather rapidly, and the fourth super, if required, may begin to fill rapidly and then fill rather slowly or stop filling.

It cannot be stressed too strongly that you keep well ahead of the honey flow in providing super space. However, when a super begins to fill slowly (indicating that the flow is tapering off or ending), do not add another super until the one on the hive is very nearly full. As the flow tapers off, make the bees fill all the available space in the supers on the hive before adding another.

Inspecting the Honey Super

So that you don't mash or roll the bees by pulling a frame out of the center of the honey super (or the brood super), remove a side frame and set it aside in an empty super, or stand it on end in a safe place. Loosen and slide the frames aside until you come to the frame you wish to remove and inspect. The bees tend to work the center frames in a super first, and then they work outward to the sides of the super. Remove and inspect as many frames in the super as you wish. Then manipulate the frames in the super, as discussed in the next section, to get all the frames drawn and filled as evenly as possible.

Manipulating the Frames in the Honey Super

Because the bees tend to work the center frames in the super first, move those partially drawn and filled frames to the side of the super and place undrawn foundation or unfilled comb in the center of the super so it can be drawn and filled faster.

Always perform this manipulation at the time you add a new honey super, particularly if the frames of the super presently on the hive were foundation. The reason: The bees may move up into the newly added super and neglect the outer frames in the super below, and these frames may not be as nicely drawn and filled if the flow slows down or ends before they are needed.

Consider These Things When Adding Honey Supers

1. Should you top super or bottom super when adding honey supers?

    There is a difference of opinion about whether it is better to top super or bottom super. Read the following discussion and decide which is better for your particular situation.

    Top supering: Top supering is adding the next honey super directly on top of the honey super or supers on the hive each time another super is added. Top supering is much easier from the standpoint of work involved. One does not have to lift the supers off and on. It is easier to check how full the last super added is without moving supers off and on. The author uses bottom supering on the first three honey supers, because they are not too heavy to handle when the third super is added. However, the fourth super is always top supered to avoid lifting full supers of honey off and on, with the fullest super to be placed back on the hive at shoulder height.

    Bottom supering: Bottom supering is accomplished by removing the honey super or supers on the hive and adding the next honey super on the queen excluder, then replacing the honey supers removed on top of the new super just added.

    Consider the following in favor of bottom supering:
    - Bees seem to draw foundation into comb more quickly if it is just above the brood chamber. The honey super moved up continues to be drawn and filled as the comb is being drawn in the new honey super.

- When bottom supering, the fullest honey super is always at the top of the hive. If for some reason you want to extract in the middle of the honey flow, just remove the topmost super, and if 95 percent of the cells are capped, you can extract. If you extract a honey super in the middle of the honey flow, you can immediately return that super to the hive and make three supers do the work of four. If you want a reason to buy an extractor and you have a number of hives, consider extracting in the middle of the honey flow saving the cost of one super, frames, and wax per hive. What you save you could spend on an extractor.
- If you are producing comb honey, bottom super so that as the supers fill they will be moved to the top of the hive to be removed when ready, before the lovely white cappings become travel stained.
- If you have foundation to be drawn into comb, bottom super during the honey flow. In several years all your supers will have drawn comb and you will want to top super, unless you want to remove and extract in the middle of the honey flow.

2. Add honey supers with foundation during the honey flow.

Bees will draw foundation into comb most rapidly during the honey flow. As long as bees have to hold nectar and have no place to put it, the wax glands of the younger bees secrete wax flakes, and comb is drawn rather quickly. There are four pairs of wax glands on the underside of the bee's abdomen. As the wax flakes are secreted and harden, the bee stabs them off with a spine on its middle leg, brings them (one at a time) to its mouth, and masticates and molds the wax flake into a tiny portion of a wax cell. Together, thousands of bees can draw comb rapidly.

When the honey flow declines, the bees are not holding nectar, and if there is foundation in a super, the bees will borrow or chew away the foundation wax to use as cappings, etc., in the hive. Sometimes there is considerable mutilation of foundation by this borrowing. It is particularly bad if the bees chew the plastic base foundation, for wax removal will leave the bare plastic. Because the bees cannot add comb to the plastic, it will be bare forever in the frames.

The bees do not seem to reuse or mutilate drawn comb to any extent, nor do they seem to reuse wax cappings if offered at this time. When the wax has been melted down to be made into foundation, it apparently changes the chemistry and the bees will borrow and reuse it.

3. Should the first honey super added be comb or foundation?

The first honey super added can be either comb or foundation. However, if you have enough supers of comb, use the comb. Item 4 strongly suggests that you save a super of comb to be used as the last super of the season to eliminate the mutilation of foundation. If using a super of comb as the first honey super would mean using a super of foundation as the last honey super of the season, use foundation for the first honey super. As mentioned earlier, add supers of foundation during the honey flow until all your supers are full of comb.

4. The last super added should always be comb.

The reason that the last super should always be comb was alluded to in items 2 and 3; namely, to keep the bees from mutilating the foundation in the last super added when the honey flow ends. The two exceptions are:
- For the beginner who has no comb available, all supers added the first year will have foundation in the frames.
- The beekeeper who forgets and uses all his or her comb too early.

Always hold back a super of comb (stored in a plastic bag with PDB moth crystals) for each hive that you have, to use as the last honey super of the season. If you follow this advice, you will not be forced into giving the hive a super of foundation that could be mutilated when the honey flow ends. The beginner who has only one honey super of comb at the end of the first season should save it to use as the last honey super next year and use foundation during the early part of the next season. Eventually all your honey supers will have drawn comb.

5. Plan ahead with honey supers: Draw them this year for use next year.

If you plan to add more hives to your apiary next year, hold back your supers of comb under moth crystals, and have your hives draw foundation into comb during the honey flow. Each hive should be able to draw two supers of foundation into comb during the honey flow. Your first super will be comb; add two supers of foundation and a super of comb at the end of the season. Next year you will be able to put packaged bees on one super of comb, saving the second super for the last super of the season. If you plan to start new hives from swarms, let them draw foundation into comb, and save the supers of comb for the last honey super of the season.

If you have a super that is half full of comb and you want to use it as a first honey super, somewhere you will have a super that is half full of foundation. Because bees work the center first, put the comb on the sides of the super and the foundation in the center of the super. The comb will be drawn out more evenly on the foundation.

6. Should you use nine frames in a ten-frame super?

The answer to the question is yes, but *only if you read and use* the following precautions for going from ten frames down to nine frames.

Caution: You should *not* start with nine frames of foundation in a honey super.

You should always start with ten frames of foundation in the honey super. When the frames have been drawn into comb, filled, and extracted, reduce the number of frames in the honey super to nine. Failure to heed this advice will result in many frames being crosscombed as they are drawn. (Crosscombing is two or more frames fastened together by comb running between the sheets of foundation.) These messy combs are never nice to work with when extracting.

Crosscombing probably results because the sheets of foundation are farther apart with nine frames, and for some reason, the bees begin drawing comb between the foundation sheets rather than along them. Crosscombing occurs much more frequently between nine frames of foundation than between ten frames of foundation. If you find crosscombing, remove the frames involved and tear the crosscombs off right down to the foundation; make the bees do it right.

More honey is stored in nine frames per super than in ten frames, which makes the combs easier to extract. Each comb is drawn a little farther out, leaving fewer low spots to scratch the cappings from manually. Metal frame spacers are available at bee supply houses that will hold nine or eight frames evenly spaced in the ten-frame super. The metal frame spacers in the super, keeping the frames evenly spaced, are handy when carrying supers of empty comb to the hive. Otherwise, you must manually space the frames in each super after setting it on the hive.

Some beekeepers even use nine frames in the brood chambers, which probably gives a little better ventilation and a little more cluster room between the frames. It also helps to eliminate some crowding before the hive swarms. The author has never tried eight frames in a ten-frame honey super, nor has he tried nine frames in a ten-frame brood chamber to verify that it is a good wintering configuration in the North.

7. Do you want more wax?

Do you want more wax to sell or make candles, etc.? Are you getting more honey than you can sell or give as gifts? Turn some of that honey into wax.

Honey/nectar is consumed in the secreting of wax flakes and in the drawing of comb. Depending on the temperature at the time the comb is being drawn, six pounds or more of honey are used for drawing one pound of wax.

You must decide how many supers you want to put into the production of wax and store supers of comb that will not be needed in plastic bags with PDB crystals. You might get 2 or 2½ supers of honey using the following method. Melt down the combs in the supers you will use, and replace starter strips of foundation in the frames.

To get more wax, obtain 5⅝-inch cut comb foundation or 8½-inch medium brood foundation with no wires. Cut the smaller foundation in half, or cut the larger foundation in thirds. Attach these starter strips with the wedge bar to the frame top as usual. Support the starter strips with support pins. As the honey flow begins, add these honey supers (with the starter strips) on the hives as you would any honey supers. As the need for honey storage space in the hive increases, the bees will hang from the starter strips and draw natural comb down to the bottom bar. The comb produced is not good for anything but melting down after it is extracted, because most of the cells are storage cells. Storage cells are about half again as large as worker cells; if used for brood rearing they would produce only drones.

The comb drawn in these supers follows the same principle as the comb drawn in a hive in a hollow tree, except that the starter strips keep the bees on track along the frame, and gravity brings the drawn comb down to the bottom bar. Add these honey supers with starter strips as required during the honey flow. You can add a super of comb for the last honey super of the season or use the honey super with starter strips.

In the fall, uncap and extract the combs that are drawn and filled. Some of the comb may break in the extractor, but it doesn't matter, because you are going to melt it down anyway. If the comb extracts well, place the frames back in the supers and give them back to the bees for several days to clean up. Then remove the supers, cut the comb out of the frames, and melt it down. Next season, add new starter strips and have the bees make new comb again. When you have all the wax you want, or when you want to go back to producing honey, buy the vertical-wired or plastic-reinforced foundation and insert it in the frames; let the bees draw it out the next year during the honey flow. You will also use the supers of comb stored earlier.

ADDING HONEY SUPERS

Once again, keep ahead of the honey flow with adequate honey storage space. It is better to have more space than too little. It is best to add one deep super or a pair of medium or shallow supers at a time and add another super in a week or two. However, if you are going to be away for several weeks during the honey flow, add extra super space all at once, if necessary, to make sure the supers will not fill completely and cause the hive to swarm.

For the first few years, keep notes each time you inspect the brood or honey supers, particularly the honey supers. In reviewing your notes at the end of the year and over several years, you will get to know when your honey flow starts, when the honey flow is heaviest, and when the flow ends. Eventually you will be able to schedule your honey supering by calendar dates and not have to check the supers, except when you add the next super.

Adding the First Honey Super and the Queen Excluder

The first honey super should be added about three weeks before the honey flow is to begin (sooner if the first honey super is to be used for swarm control by moving frames of brood above the queen excluder). The reasons for adding the honey super well in advance of the honey flow are: (1) to give extra room for population expansion and (2) to have storage space available whenever the incoming nectar exceeds brood rearing needs.

Do not worry if there is almost no activity in the first honey super for a few weeks. The bees will not move into the honey super through the queen excluder until they need to store excess nectar. When that time comes, the bees will begin to move into the honey super to clean up the comb and cells; then nectar will begin to be stored and bees will be found in large numbers. Many beekeepers blame the queen excluder for the bees not moving up, but the reason is that there is no excess nectar to be stored. Have faith in the bees and they will do what is necessary in their own time.

Remember:
- Follow the rule of thumb for adding honey supers.
- Store one super of comb to add as the last honey super.
- Make notes of your honey super inspections until you become familiar with the honey flow in your area.

If you are adding the honey super and queen excluder as a swarm control method, see Unit 15 for adding the first honey super and moving frames of brood above the queen excluder. Otherwise, follow these steps.

Equipment needed:
- [ ] Hive tool
- [ ] Smoker or hive bomb
- [ ] Frame grip
- [ ] Queen excluder
- [ ] Honey super with nine or ten frames of comb or ten frames of foundation
- [ ] A pair of two-by-fours about 18″ long

Step-by-step directions:
- ■ **Dress properly.**
- ■ **Light the smoker** or have hive bomb handy.
- ■ **Enter the hive.** You may not want to smoke the hive entrance.
- ■ **Inspect the brood chamber** (optional). Be sure to inspect the brood chamber when you add the second honey super. (See Unit 12.)
- ☐ Smoke the frame tops of the upper brood chamber. Place the queen excluder on the upper brood chamber with the smooth side of the excluder up.
- ☐ Set the honey super on the queen excluder. Space the frames evenly in the super. Add one super of any size for the first honey super; however, there is no reason why you cannot add a pair of medium or shallow supers for the first honey super. Just keep in mind that a single shallow super will fill more quickly than a deep super, so you will have to check it more often. If you cannot depend on yourself to check the shallow super more often, then by all means add a pair. Close the hive.

Inspecting the Honey Super or Pair of Honey Supers

Equipment needed:
- ☐ Hive tool
- ☐ Smoker or hive bomb
- ☐ Frame grip
- ☐ Empty super

Step-by-step directions:
- ■ **Dress properly.**
- ■ **Light the smoker** or have hive bomb handy.
- ■ **Enter the hive.** You may not want to smoke the hive entrance.

A. Inspecting a single honey super of any size.
- ☐ Remove one or two frames from one side of the super, and place them in the empty super or stand them on end nearby.
- ☐ Loosen and move the frames, one by one, to the side of the super. Check the middle four frames. If any of the middle frames are partially drawn or filled, move them to the sides of the super as follows. Remove one or two frames of undrawn foundation or empty comb from each side of the honey super, and place them in the empty super or stand them on end nearby. Lift out the partially drawn or filled combs from the center of the super, and place one or two at each side of the honey super. Replace the undrawn foundation or empty combs in the center of the honey super.

The reason for this manipulation is to keep undrawn foundation or empty comb in the center of the honey super, where the bees seem to work the hardest. It also gives the beekeeper a chance to see if any crosscombing is occurring between the foundations of adjoining frames. It gives the beekeeper a feel for when seven frames are partially drawn and filled and when it is time to add the next honey super. Drawn comb does not have to be manipulated as outlined above, but it may help the beginner in deciding when to add the next honey super.

- [ ] Make notes on a calendar or in a notebook each time you inspect the honey supers. Make a note each time a honey super is added.

B. Inspecting medium or shallow supers added in pairs.
- [ ] Place the inverted hive top on the ground nearby.
- [ ] Lift the upper of the two honey supers and puff in some smoke. Wait one or two minutes, and then set the upper honey super on the inverted hive top.
- [ ] Manipulate the frames in the honey super on the hive as directed previously for inspecting a single super. Continue to inspect and manipulate frames about once each week.
- [ ] When upon inspection six or seven of the frames in the lower honey super are partially drawn and filled, smoke between the honey super and queen excluder and wait one or two minutes. Set the honey super off onto the pair of two-by-fours.
- [ ] Place the honey super that is on the inverted hive top (from the second step in this section) on the queen excluder. Manipulate the frames if any of them have begun to be drawn and are partially filled.
- [ ] Place the honey super now on the two-by-fours onto the honey super on the hive. In these last three steps you have reversed the pair of honey supers onto the hive.
- [ ] Close the hive. Continue to inspect and manipulate frames until it is time to add another pair of honey supers. During the height of the honey flow, pairs of honey supers may fill together. In this case, manipulate the frames in both supers at each inspection.

Adding the Second Honey Super or Pair of Honey Supers

Equipment needed:
- [ ] Hive tool
- [ ] Smoker or hive bomb
- [ ] Frame grip

- ☐ Empty super
- ☐ Pair of two-by-fours
- ☐ Super or supers to be added (They should be aired for two days if they were stored with PDB.)

A. Top supering.

Step-by-step directions:
- ■ **Dress properly.**
- ■ **Light the smoker** or have hive bomb handy.
- ■ **Enter the hive.** You need not smoke the hive entrance. Place the inverted hive top on the ground nearby. Smoke between all honey supers, wait one minute, and set them on the inverted hive top.
- ■ **Inspect the brood chamber.** (See Unit 12.) This is an important inspection, because you want to be sure there is a queen in the hive as you go into the rest of the honey flow. It is also important because you may find that the hive swarmed, unknown to you, and the virgin queen may not have come back from her mating flight, which would leave the hive queenless. If you find the hive is queenless, order a new mated queen and re-queen as directed in Unit 13. Check for laying workers and wax moth infestation before ordering a new queen.
- ☐ After the inspection is completed, replace the queen excluder. Replace each honey super on the hive, making a final manipulation of the frames in each super as they are replaced.
- ☐ Place the empty honey super or pair of supers on the honey supers presently on the hive.
- ☐ Close the hive. Continue to inspect the honey supers at weekly intervals and manipulate the frames in the supers as directed earlier. Remember, nectar is coming in faster every day, so do not wait too long between inspections. Make notes on each inspection, including dates.

B. Bottom supering.

Step-by-step directions:
- ■ **Dress properly.**
- ■ **Light the smoker** or have hive bomb handy.
- ■ **Enter the hive.** You need not smoke the hive entrance. Place the inverted hive top on the ground nearby. Smoke between all honey supers, wait one minute, and set them on the inverted hive top.
- ■ **Inspect the brood chamber.** (See Unit 12.) This is an important inspection, because you want to be sure there is a queen in the hive as you go into the rest of the honey flow. It is also important because

you may find that the hive swarmed, unknown to you, and the virgin queen may not have come back from her mating flight, which would leave the hive queenless. If you find the hive is queenless, order a new mated queen and re-queen as directed in Unit 13. Check for laying workers and wax moth infestation before ordering a new queen.
- [ ] After the inspection is completed, replace the queen excluder, and add the empty honey super or supers on the queen excluder.
- [ ] Set the honey supers that were on the inverted hive top on the empty super or supers. Manipulate the frames in these honey supers as you return them to the hive. These last two steps are the bottom supering of the hive. Inspect the newly added supers as well as the first supers added at weekly intervals. Manipulate the frames as directed earlier. Make notes after each inspection.

Adding the Third Honey Super or Pair of Honey Supers

Equipment needed:
- [ ] Hive tool
- [ ] Smoker or hive bomb
- [ ] Frame grip
- [ ] Empty super
- [ ] Super or supers to be added (They should be aired for two days if they were stored with PDB.)
- [ ] Pair of two-by-fours

A. Top supering.

Step-by-step directions:
- ■ **Dress properly.**
- ■ **Light the smoker** or have hive bomb handy.
- ■ **Enter the hive.** You need not smoke the hive entrance.
- ■ **Inspect the brood chamber** (optional). It is not necessary to inspect the brood chambers at this time, unless the hive does not seem to be prospering. If you have other hives and one does not seem to be keeping up with the others in honey production, you may want to have a look. (See Unit 12.)
- [ ] Inspect the last super or pair of supers added. Manipulate the frames in the honey super as directed earlier.
- [ ] Place the empty super or pair of supers on the supers now on the hive. Close the hive. Make notes on all honey inspections. Inspect the honey super or supers at weekly intervals.

B. Bottom supering.

Step-by-step directions:
- ■ **Dress properly.**
- ■ **Light the smoker** or have hive bomb handy.
- ■ **Enter the hive.** You need not smoke the hive entrance. Place the inverted hive top on the ground nearby.
- ☐ Smoke between all honey supers on the hive (down to and including the queen excluder). Wait one or two minutes. Set all the honey supers on the inverted hive top.

  If you want to inspect the brood chambers, you can do so at this time by smoking over the queen excluder and removing it, then following the steps in Unit 12 to inspect the brood chambers. Continue with the next step after you inspect the brood chambers.
- ☐ Add the empty honey super or supers on top of the queen excluder. Replace the supers on the inverted hive top back onto the empty super or pair of supers. Inspect the supers as they are replaced and manipulate the frames, if necessary. If, when you replace the first super or pair of supers, you find them filled and at least 95 percent capped, you can begin to remove those filled supers and extract, or if producing comb honey, you can remove the supers and begin to process and sell the comb honey.
- ☐ Close the hive. Inspect the honey supers at weekly intervals. At some point in time, if the topmost supers have not been removed and extracted, they will become too full and too heavy to make the hive fun to work. When this time arrives, do not inspect the bottom supers anymore, but top super with all additional honey supers that you place on the hive from now on.

What to Expect at This Time in the Honey Flow

This section is a quick summary of what you have added thus far and what you can expect to be happening in those honey supers added up till now. It suggests what you can do about removing honey supers if you want to extract in mid-honey flow and about removing supers of comb honey as they become fully capped.

1. In general, you have now added your third deep super or your fifth or sixth medium or shallow super. As a rough reference, in the South the time is about May 1 to May 15. In the (far) North it is about July 1 to July 15. The honey flow is usually at its peak now and may hold for several more weeks, then begin to taper off over the following three weeks or so. After that, the honey flow can be considered over.

In the North: When the honey flow is over, be prepared to remove the honey supers, extract the honey, clean up and store the extracted honey supers, and begin to medicate/feed the hive in preparation for winter. Don't wait too long, or the weather may become too cold for the bees to participate in these activities.

In the South: You will most likely have several more months of nice weather after the honey flow ends. You may extract when the flow is over or wait for another couple of months. Remember that the bees can take better care of the honey on the hive than you can off the hive.

2. If you have been inspecting and manipulating the frames in your honey supers, you will notice at this time that the first super (or pair of supers) added is fully capped, or nearly so. If this is the case, remove that super or those supers to extract, and then return the super or supers to the hive. If you are producing comb honey for sale, you should remove the full supers as they become fully capped and prepare them for sale.

If you have been top supering, the full super or supers are at the bottom of the stack of honey supers. If you have been bottom supering, the full super or supers are at the top of the stack and can be removed easily from that position.

Why Remove Supers and Extract in Mid-Flow?

You may want to extract in mid-flow under the following circumstances.

- You have sold all your honey and your customers are clamoring for more of your good honey as soon as you can get it.
- You have no comb to put on your hive as the last honey super of the season. You can extract one of the supers and return it to the hive as comb.
- You own an extractor or want a good reason to buy one. You can remove a super from each hive, extract it, and return it to the hive. In this case, every super you return to the hive is a super you do not have to buy. If you have ten hives, that will save at least enough to buy a small extractor. It also means less extracting to do in the fall.
- If you have the time and the inclination, extract twice during the flow: once when the first super added is full and capped, and again when the second super added is full and capped.

CAUTION: If you extract during the honey flow, do not extract any frames that are less than 90 to 95 percent capped. Extract only those frames that are almost entirely capped, and return the others to the hive (along with the extracted comb) to be finished off and capped. A small amount of green (watery) honey can cause all your honey to ferment and become inedible.

Adding the Fourth and Following Supers

In areas where the honey flow may be very strong, such as most northern areas where large acreages of nectar-bearing flora exist naturally or are cultivated, you may find that five or more supers may be required for honey stores. Most of us are not that lucky and will find that four deep honey supers are all or more than we need. Four deep supers equal approximately six medium supers or approximately eight shallow supers.

If you find that you require five or six honey supers, save back two supers of comb each year to add by top supering. As more and more supers are added to the hive, top supering becomes more advantageous.

Equipment needed:
- ☐ Hive tool
- ☐ Smoker or hive bomb
- ☐ Frame grip
- ☐ Super or supers of empty comb to be added (They should be aired for two days if stored with PDB.)
- ☐ Pair of two-by-fours

Step-by-step directions:
- ☐ Top super from now on.
- ■ **Dress properly.**
- ■ **Light the smoker** or have hive bomb handy.
- ■ **Enter the hive.** You need not smoke the hive entrance.
- ☐ Now is the time to add that super of empty comb you have been saving. Place the super of comb on top of the supers that are on the hive. If you will be using five or six supers, they will be added as in the previous steps, by placing them on top of the topmost super on the hive (top supering). As long as the flow remains strong in your area, continue to add supers of foundation, if necessary, and save your comb for the last one or two honey supers added.
- ☐ Close the hive. Continue to inspect the honey super at weekly intervals, if possible. Make notes after each honey super inspection.

About the Notes Made After Each Brood and Honey Super Inspection

When you review your notes of brood and honey super inspections for last year, this year, and (later on) the next year, you will begin to get a feel for the honey flow in your area. You should have a good idea of when the flow starts, when it peaks, and when it ends. Using this information, plan a schedule for adding supers by calendar dates. This does not mean that you should give up inspecting, but it can be less frequent, mostly at the time you add the next super.

For example, in the San Jose, California, area, the author adds honey supers, in the following calendar sequence. The dates may be altered somewwhat, depending on the rainfall.

1st honey super	March 15-20	Super fills very slowly.
2nd honey super	April 10-15	Honey flow is getting strong.
3rd honey super	May 1	Honey flow is at or near its peak.
4th honey super	June 1	Honey flow is beginning to weaken and will end around July 1-15.

The author does not begin to extract until about September 1 when the weather is cooler and the massive population of bees has dwindled almost to winter strength. In this area, the bees are more aggressive after the honey flow ends. This is probably because the population stays strong while at the same time there is no honey flow. When the beekeeper goes to work the bees, most of the older field bees are in the hive and unhappy that they cannot be gathering nectar.

UNIT 19

Removing Honey Supers

The time will come for removing the honey supers and extracting the golden treasure stored therein. The main problem at this time is removing the bees from the supers so they can be stored in the garage or house until all supers are removed and you can begin to extract.

One other problem must be considered as you remove the supers and store them for extracting: The wax moth eggs laid in the supers while on the hive (they are in almost every hive in the world) begin to hatch, and the larvae will feed and grow in those supers and frames of honey. Unit 20 covers storing comb honey so that it will not become infested with wax moths. You will have about two weeks in which to extract or store your comb honey after removing it from the hive before the wax moths begin to infest your combs of honey.

Bees can be removed from the honey supers using the following methods, which are discussed in detail in this unit.

- Brushing the bees from each frame in the super.
- Using a bee escape in the oval hole in the inner cover.
- Using a fume board and a chemical scent to push the bees out of the super.
- Using a high-velocity blower to blow the bees from the frames after the super has been removed from the hive.

BRUSHING BEES FROM FRAMES OF CAPPED COMBS

Brushing the bees from each frame is probably the poorest way to remove bees from the supers of honey. It is suggested that you do not try it, but if you must, do dress for the fight for the honey stores. You have my sympathy.

Your task will not be quite so hard if you shake the frame to dislodge most of the bees, then brush the rest of the bees off the combs of capped honey. Before and during the honey flow, shaking then brushing the bees from the frames of capped honey does not seem to antagonize the bees nearly so much as it does when the honey flow is over and the older field bees are not busy gathering nectar. You will remember that unless the scout bees find nectar and dance exitedly, not many of the field bees will leave the hive.

Equipment needed:
- [ ] Hive tool
- [ ] Smoker or hive bomb
- [ ] Frame grip
- [ ] Bee brush
- [ ] Empty super
- [ ] Damp dish towel

Step-by-step directions:
- ■ **Dress properly.**
- ■ **Light the smoker** or have hive bomb handy.
- ■ **Enter the hive.** Place the inverted hive top on the ground nearby. Set the empty super *inside* the inverted hive top so the bees cannot get back onto the combs from underneath.
- [ ] Smoke the frame tops of the topmost honey super. Remove a frame from the super. Step toward the front of the hive.
- [ ] Grasp each end of the frame top with three fingers under the end of the top bar and a thumb on top of the end of the top bar. Push the frame rapidly downward and slightly outward about one foot, and then stop the descent abruptly. Most of the bees will fall from the frame of comb.
- [ ] Holding one end of the frame top bar with one hand, brush off the bees still on the comb (brush them toward the front of the hive). Brush both sides and the bottom.
- [ ] Place the frame, now free of bees, in the super that is *inside* the inverted hive top. Drape the damp dish towel over the super to keep the bees from getting back on the comb from above. If you find that robbing is occurring in the open super from which you are brushing the bees, you may want to drape another damp dish towel over the super on the hive or replace the inner cover on the super.
- [ ] Repeat the previous steps for each frame in the super and for as many supers as you are going to remove. With each comb free of bees, turn back the dish towel, place the frame in the empty super, and pull the dish towel back over the super.
- [ ] When all frames from the super have been shaken, brushed, and placed in the empty super inside the inverted hive top, carry the

super to the garage or other safe place where the bees cannot get to the honey. If you are going to remove another super from the hive, brush the bees from the now-empty honey super on the hive, and set it inside the inverted hive top. Repeat all the previous steps.

If you are going to remove only one super of honey from the hive, remove the now-empty super on the hive, and replace the inner cover and hive top.

If you plan to remove all supers, continue with the previous steps until you reach the queen excluder, remove the last empty super, remove the queen excluder, and replace the inner cover and hive top.

CLEARING BEES FROM HONEY SUPERS WITH THE BEE ESCAPE

The bee escape was invented about the time of the Civil War, after the movable frame hives were developed, to aid in removing bees from frames of comb without brushing. The Porter Bee Escape replaced the goose wing brush that probably was used prior to that time. Most honey was produced in basswood section boxes for many years, because extractors were not in common use or were considered an expensive luxury.

To this day, most complete ten-frame hive kits include an inner cover with an oval hole. The bee escape fits in the oval hole. When the bee escape is in place and the inner cover is slipped under a super full of honey and bees, the bee escape acts as a one-way gate. The bees move out of the hole in the bee escape, press through a pair of light springs, and cannot return. Usually within twenty-four hours, most of the bees will have been removed from the honey super.

The bee escape works best when the evenings are cool. In northern areas, that would be after the honey flow is over in the fall. In southern areas, the days are warm—sometimes even warmer—after the honey flow than during the honey flow. In those areas where the temperature reaches 90° F or more at the time you will be removing the honey supers, consider using a screened inner cover or a method of removing the bees other than the bee escapes. You can make a modified inner cover that is partially screened so the bees below can continue to ventilate or cool the super from which the bees are being removed through the bee escape.

On hot days, if you are not using the screened inner cover, put the bee escape and inner cover on the hive in the later part of the afternoon. This will give most of the bees a chance to leave the honey super during the cooler night. Remember, a super closed off, except for the hole in the bee escape, can reach a fairly high temperature during the heat of the day (like a closed car).

Equipment needed:
- ☐ Hive tool
- ☐ Smoker or hive bomb
- ☐ Frame grip
- ☐ Bee escape for each hive to be worked
- ☐ Pair of two-by-fours
- ☐ Covered can or pan

Step-by-step directions:
- ■ **Dress properly.**
- ■ **Light the smoker** or have hive bomb handy.
- ☐ In mid-honey flow, if the fullest super or completed comb honey super is not on top of the hive, you will have to shuffle the supers around to get them in that position. If the honey flow is over and all supers are to be removed, the order of the supers makes no difference.
- ☐ If the inner cover is not on the hive, do not enter the hive, but proceed to the next step. If the inner cover is on the hive, ■ **Enter the hive,** remove the inner cover, and replace the hive top.
- ☐ With the hive tool, lift the topmost super about one inch and smoke between the two topmost supers. Wait one or two minutes. While waiting, use the hive tool to loosen the super on all sides so it can be removed from the hive or tilted up to insert the inner cover with the bee escape in place.
- ☐ Choose method A or B for inserting the inner cover under the full super of honey.
    A. Remove the honey super from the hive and set it on the inverted hive top or across the pair of two-by-fours. Smoke the top of the exposed super on the hive.
- ☐ Scrape any burr comb from the frame tops of the super on the hive, and put the scrapings in the covered pan. Place the inner cover on that super and insert the bee escape. This will keep the bees that are down below in the hive.

    Now, scrape any burr comb from the bottom bars of the super that is on the inverted hive top or across the two-by-fours. Place the scraps of comb in the covered pan or can to keep the area clean of honey-covered scraps removed from between the honey supers, which will lessen the chance of robbing.
- ☐ Replace the just-scraped super on the inner cover with the bee escape in place. Go to step C.
    B. Slide the loosened super toward you about one inch (you should be behind the hive), so that when you tilt the super up it will not slide off the front of the hive. (See Figure 19-1.)

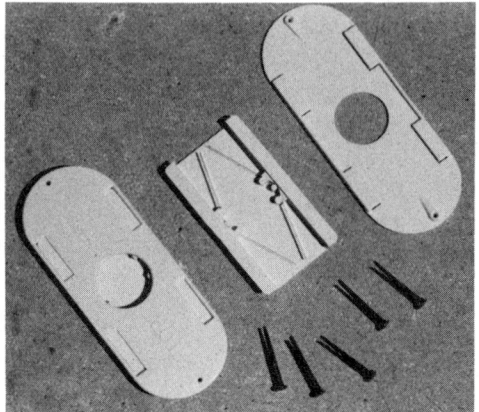

Figure 19-1, Inserting the Inner Cover and Bee Escape. Left: The bee escape as a unit and a bee escape dismantled to show the wire springs inside. Also shown are several support pins. Below: One method of inserting the inner cover and bee escape under a full super of honey. Notice that the honey super was pulled forward about one inch to provide a fulcrum point so the honey super will not slide off the hive when it is pivoted up to near vertical.

- [ ] Have the inner cover and bee escape where they can be reached easily. Tilt the honey super up about 90°. Smoke the top bars of the lower super and the bottom bars of the tilted super. Scrape any burr comb from the lower super, and place it in the covered pan or can.
- [ ] Place the inner cover on the lower super, pushing one end against the tilted super. Insert the bee escape in the oval hole of the inner cover.
- [ ] Scrape any burr comb from the bottom bars of the upper super, and place the scrapings in the covered pan or can. Lower the tilted super down onto the inner cover, checking to make sure the bee escape is in place.

- [ ] Pull the honey super toward you until it is squarely on the inner cover. Push the inner cover and super forward until it is squarely on the hive. Continue with step C.
    C. After about thirty minutes, check the hive to see if there are bees entering or exiting from any nicks or cracks on the edges of the super above the inner cover. Does the super fit snugly on the inner cover so bees cannot enter or exit around the inner cover? Does the top fit snugly, or does it have a warp so that bees can enter and exit from under the hive top?
      Close any nicks or cracks with a rag or masking tape. If the top does not fit properly, you may have to cut a 16¼-inch by 20-inch rectangle of masonite or plywood to cover the super. If bees can come and go through any opening in the super on top of the inner cover with the bee escape in place, the honey stores will be robbed out; when you come back to remove the honey super, it may be empty of honey or still full of bees.
- [ ] *Wait twenty-four hours.*
- [■] **Dress properly.**
- [■] **Light the smoker** or have hive bomb handy.
- [ ] Gently lift the hive top to see if the bees are out of the super. If most of the bees are out of the super, go to the next step. If the super still seems to be full of bees, the next section describes some problems you may encounter using the bee escape and tells you how to cope. In this case, it is most likely that you will have to brush the bees or find the cause of the problem and have another go at getting the bees out of the super by repeating the previous steps.
- [ ] With the hive tool, loosen the super from the inner cover and carry it to the garage, house, or other safe area where the bees cannot reach it.
- [ ] If you are removing only one super, remove the bee escape from the inner cover and replace the hive top; otherwise, go to the next step.
- [ ] If you are removing all supers above the queen excluder, repeat the steps above, removing one super every twenty-four or forty-eight hours until all supers above the queen excluder are removed.

Some Problems You May Encounter Using the Bee Escape

If the bees are not almost completely removed from the honey super in twenty-four to forty-eight hours, there may be some problem with the bee escape, the super may have holes, the top may fit poorly, or the weather may be quite hot.

Possible problems that could keep the bees from leaving a honey super through the bee escape could be any of the following.

1. Brood may be in the honey super, which may occur if no queen excluder is used. The nurse bees and the very young bees do not leave the super, and new bees emerge every day, so the super will not clear. You will have to brush the bees from the combs of honey.
2. The spring openings of the bee escape may be too far apart. If the openings are more than about ⅛ inch, the bees may be able to get back through the springs into the honey super.
3. The spring openings may be too tightly closed, or they may have been glued fast with propolis. In either case, the bees cannot force the springs apart enough to exit the honey super.
4. The opening of the bee escape from above or below may be closed by wax or burr comb. If the burr comb is not removed before adding the bee escape, it may mash into the openings and stop all traffic completely.
5. Days may be hot or you may be adding the bee escape too early in the day. If the bee escape is added too early on hot days, the bees may begin to die, and other bees try to pull them through the springs and clog up all exits.
6. In a heavily populated hive, after the bees are removed from several supers there may be no room for more bees below the honey super. If this appears to be the problem, add an empty super with frames of comb or foundation under the last honey super to be removed to make room for the bees leaving the honey super through the bee escape.
7. Nicks, cracks, and poorly fitting tops or inner covers have been discussed in the steps above. These problems could lead to robbing and should be carefully watched for.

If the bee escape does not work, you may have to resort to brushing the bees from the frames as directed earlier. Or you may want to smoke under the super, set the honey super aside, and inspect the bee escape and super for the problems mentioned above. When the problem has been resolved, set the honey super on the inner cover, with the bee escape in place, and wait another twenty-four to forty-eight hours before removing the honey super.

USING A FUME PAD AND A CHEMICAL BEE-REMOVING LIQUID

Bees tend to move away from the offending odor of certain chemicals. The advantage of the fume pad sprinkled with a chemical is that you can remove all the supers one after the other and be done with the job in a single day. The disadvantage of the chemical removers is that they leave more bees in each super than do properly working bee escapes.

The first chemical used commercially to remove bees from the supers was carbolic acid. Carbolic acid is presently banned by the Food and Drug Administration. The chemicals now approved for use in chemically removing bees from honey supers are Bee-Go® and benzaldehyde. Bee-Go is a liquid that smells rather terrible to many beekeepers, as it must to the bees. Benzaldehyde is a liquid that has a smell similar to oil of almond. Both do a good job, so do not think that if one smells worse it should work better.

Remember that any chemical remover should not be left on the hive too long, for it could taint the comb and honey and drive the bees from the brood chambers, possibly starving, overheating, or chilling the brood. The fume pad or fume board is used to keep the chemical from coming in contact with the burr combs or wood of the frame tops.

The fume board (or fume pad) is a two-inch rim the size of the super (16¼ by 20 inches). The top is usually metal or masonite with a flannel pad on the underside of the top. The chemical is sprinkled on the pad when the fume board is used. After the hive top and inner cover have been removed, the fume board is placed on top of the upper honey super on the hive. The sun heats the top of the fume board, the chemical evaporates slowly, and the gas drifts down through the super, driving most of the bees out of the super. (See Figure 19-2.)

Equipment needed:
- [ ] Hive tool
- [ ] Smoker or hive bomb
- [ ] Frame grip
- [ ] Bee brush
- [ ] Fume board
- [ ] Chemical bee remover (Bee-Go or benzaldehyde)

Step-by-step directions:
- [ ] A sunny day is necessary.
- [x] **Dress properly.**
- [x] **Light the smoker** or have hive bomb handy.
- [x] **Enter the hive.** You need not smoke the hive entrance.
- [ ] Sprinkle about one tablespoon of the liquid chemical remover on the cloth of the fume board.
- [ ] Place the fume board on top of the uppermost honey super on the hive. Wait about ten minutes—no longer than twenty minutes at any time. Remove the fume board and stand it nearby.
- [ ] Remove the honey super and carry it to the garage or other bee-free area.

Figure 19-2, Fume Board and Benzaldehyde Chemical Bee Remover. The fume board is shown on the topmost super on the hive, where it will set for about ten minutes. Then the super will be removed, with whatever bees remain, for extracting. The fume board is then placed on the next honey super, and the process is repeated until all honey supers are removed. Leaning against the hive is a fume board showing the flannel pad under the metal cover. On the leaning fume board pad is a quart can of the chemical bee remover, benzaldehyde.

- ☐ Replace the fume board on the topmost honey super on the hive and repeat all the previous steps. *Do not* add more of the chemical to the pad each time. One application of the chemical should do for two or three supers.
- ☐ If a number of bees remain in the supers, you may want to brush the bees from the frames; otherwise, stack the supers in the garage. The excess of bees in the supers will fly to the window. If you open the window about half an hour before dark, most of the bees will fly out. Do not forget to close the open window, or the bees will steal away the honey the next day.
- ☐ If you think crowding may occur as you force the bees into the brood supers, add a super with foundation or comb under the last honey super before removing that last honey super.
- ☐ When all the honey supers have been removed, replace the inner cover and the hive top.

USING A HIGH-VELOCITY BLOWER

The bee blower has the advantage of permitting the beekeeper to remove all the honey supers one after the other on the same day, leaving almost no bees in the supers. Its disadvantage is that it is quite expensive. (See Figure 19-3.)

To remove the bees, the blower must move a large volume of air at a high velocity. A vacuum cleaner or shop vacuum does not have the velocity to tear the bees out and blast them away. The low-velocity blowers blow the bees out one side, and they come back in as the blower moves across the super. High velocity is important.

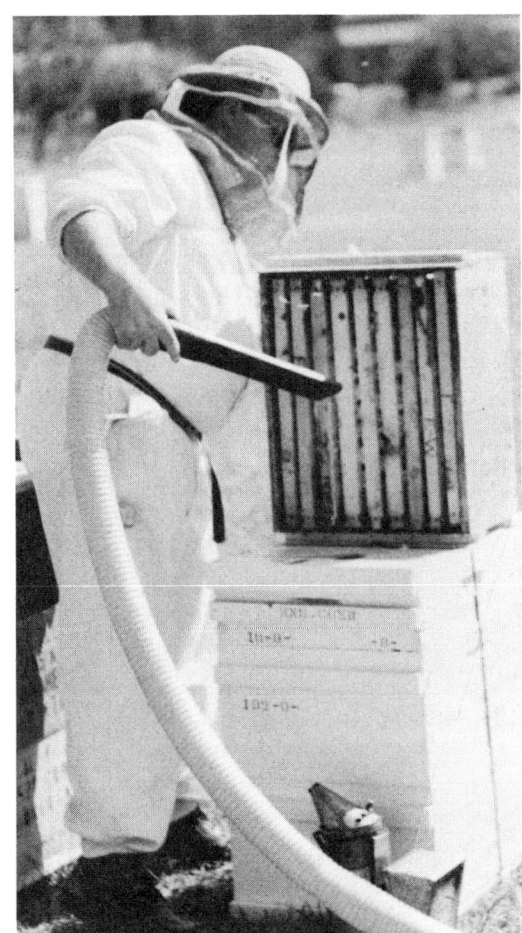

Figure 19-3, A Bee Blower in Action. Shown here is the hose and nozzle attached to a high-velocity blower. The nozzle is run along the frame tops or bottoms, blowing bees from between the frames and super sides. This virtually tears the bees loose and flings them from the frames. Once the bees have been blown out in this manner, they return to the hive.

Some of the yard blowers, the type for blowing leaves from the yard, may have enough velocity to do the job and may not be too expensive. If you can use a blower around the yard as well as for blowing bees from the supers, it might be worth considering as an investment. Look for a high-velocity machine. Check the bee supply catalogs for information on the specifications of the commercial bee blower and the chutes for holding the super of honey while blowing out the bees. You may be able to make or buy the chute if you can find a suitable yard blower. When you have a good blower, follow these steps.

Equipment needed:
- ☐ Hive tool
- ☐ Smoker or hive bomb
- ☐ Empty super or bench on which to set the honey supers for blowing
- ☐ Blower

Step-by-step directions:
- ◼ **Dress properly.**
- ◼ **Light the smoker** or have the hive bomb handy.
- ◼ **Enter the hive.**
- ☐ Remove the topmost super on the hive. Stand it on end on the empty super or bench that is standing several feet in front of and facing the hive. Replace the hive top to avoid possible robbing at the hive.
- ☐ Start the blower and blow up and down between each set of frames and between the super wall and the first frame on each side. The bees should be blown out toward the front of the hive.
- ☐ When the super has been cleared of bees, shut off the blower and remove the super to the garage or to an area that is bee-free.
- ☐ You may remove one, two, or all honey supers at any time. If you are going to remove all supers, repeat the previous steps until all honey supers have the bees blown from them. Add a super of comb or foundation if you feel excessive crowding may occur in the brood area after removing all the honey supers.
- ☐ Remove and store the queen excluder. Replace the inner cover and hive top.

UNIT 20

Processing and Storing Comb Honey

When shallow supers of comb honey in frames, section supers of comb honey in basswood boxes, or round plastic rings of comb honey are removed from the hive, two things begin to happen.

1. The honey begins to granulate or crystallize.
2. Wax moth eggs, laid on the combs and frames before the super was removed, will begin to hatch within the next few days. The larvae will feed on the propolis on the frames or comb containers, leaving behind trails of webbing and excrement.

There is a simple way to retard granulation for many months and to stop the depredation caused by the wax moth larvae. Freezing the frames of comb, honey section boxes, and plastic rings of comb in your freezer will kill the eggs and larvae of the wax moth. Storing comb honey in the freezer will cause the honey to become so thick and stiff that granulation cannot take place readily. (Jars of liquid honey also can be stored in the freezer to retard granulation.)

If you plan to sell or store a large quantity of comb honey of the sorts mentioned above, you might consider getting an old (or new) freezer for honey storage. An alternate method for a shorter storage period might be to obtain an older, working refrigerator that can be set to freezing (32° F). Honey will not granulate as quickly at a constant lower temperature and wax moth eggs will not hatch at a temperature below about 45° F. Honey tends to granulate most rapidly at a temperature of about 57° F. If the daily temperature in the storage area fluctuates above and below that temperature, the honey seems to granulate quite rapidly.

In the North: Wax moth may be a problem during the summer as you remove comb honey. When the temperature drops to and stays below 45° F, the wax moth problems will disappear, and you may store your frames of comb honey, boxes, and rings in an enclosed, vermin-free, unheated area during the winter. Extracted combs also can be stored in the same manner. The storage area should have some sort of ventilation so that the combs will not mold. Sometimes mold appears on the cappings of the comb when the comb is stored in a closed, unventilated area. Remember that as the temperature warms in the spring, the honey in the comb may granulate unless it is placed in the freezer.

In the South: Wax moths are a problem year round. Comb honey left in the open will become infested and destroyed by wax moths. If comb honey is to be left in the open for any length of time, it must first be stored in the freezer for several days to kill the eggs and larvae, then packaged in such a manner that the adult wax moth cannot get to it to lay more eggs. Honey in the comb at room temperature generally will granulate rather rapidly, so keep any comb honey in the freezer until needed for sale or personal use. Mold can also occur on combs in the South if comb honey is stored in a humid area with little or no ventilation.

PROCESSING SECTION COMB HONEY IN BASSWOOD BOXES

- ☐ Remove the section comb boxes of capped comb from the section holders. To clean up the stained areas or to remove propolis, scrape (do not sand). Three sides of the boxes usually will be clean, but the tops and edges may need to be cleaned up a bit.
- ☐ Carefully place the section boxes of comb in the freezer, and leave them for several days. Remove them and allow any condensation to dry completely, using a fan for faster and more complete drying.
- ☐ Place the section comb boxes in cartons made for storing and selling section comb honey. Cartons are available from bee supply houses. The cartons are tight enough that adult wax moths cannot enter while they are stored.
- ☐ Store the cartons of section comb honey in a dry, well-ventilated area. Remember that the comb honey will granulate less quickly if stored in a freezer or refrigerator.

Alternate method if you have freezer storage space:
- ☐ Remove section comb boxes from the section holders, and scrape the stained areas.
- ☐ Place the section boxes in cartons, and store them in the freezer until needed for sale or personal use. If the comb sections are for your own use, you can stack the section boxes carefully in the freezer without the cartons.

PROCESSING ROUND PLASTIC RINGS OF COMB HONEY

- [ ] Place the frames containing the plastic rings of comb in the freezer for several days. Remove and dry them completely.
- [ ] Remove the rings of comb honey from the frames. Scrape any soiled areas on the plastic rings or edges.
- [ ] Place the plastic covers (one clear and one opaque or both clear) on each side of the round ring, and secure them in place with the long pressure-sensitive label made for the plastic rings of comb.
- [ ] Store the packaged rings in a dry, well-ventilated area, or stack them back in the freezer until needed for sale or personal use.

PROCESSING COMB HONEY CUT FOR PLASTIC BOXES

Comb honey in the frame can be sold by the frame of comb honey, or it can be cut from the frames and placed in square plastic boxes and sold as cut comb honey. Cut comb honey is usually produced in the shallow supers. Comb honey for sale in the frame can be in either medium or shallow frames.

- [ ] Place the frames of comb in the freezer for several days. Remove and dry them completely. Use a fan to hasten the drying.
- [ ] Place the frame or frames on an oven rack or wire rack over a tray to catch the drips when the honey is cut from the frames.
- [ ] With a sharp paring knife or a knife with a thin blade, cut around the edges of the frame and lift the frame off the comb. Place the frame in a bucket to catch the drips.
- [ ] Cut the long rectangle of honey into four sections, each 4 to 4⅛ inches long. If shallow frames were used, the segments cut will fit into the plastic comb boxes that can be obtained from bee supply houses. If medium frames were used, you must cut the comb into 4- or 4⅛-inch squares to fit in the plastic comb boxes. A foil box with a plastic lid is also available for comb honey. The foil box accepts comb cut into 3- by 4-inch sections.
- [ ] After cutting the comb into sections, space them apart slightly, and let them drain overnight. The cut cells will drip into the tray, and there will be little honey mess in the plastic box to interfere with the sale of a lovely comb of honey.
- [ ] To get the comb in the plastic box, place the box over the section of honey, and squeeze the box gently while turning the box over. You can slide a spatula carefully under the comb, placing the box over the section and turning everything over.

- [ ] Place the lid on the box, and secure the lid to the box with short strips of transparent tape. Store the boxes of comb in a dry, well-ventilated area, or stack them in the freezer until needed.
- [ ] If the comb is for your own use, you may want to store the frames in the freezer. If you are in the North, you may store them in a dry, ventilated, unheated area. Bring the frame in when needed, let it warm to room temperature, and then cut it into segments. Using a spatula, lift the comb off the rack, and store it in sandwich or zip-lock bags or plastic boxes.

PROCESSING COMB HONEY IN THE FRAME

- [ ] If you are going to sell a large number of combs in the frame, you will probably want to start with new frames and foundation each year. This way you will not have quite so much scraping and clean-up work to do before you sell the frame of comb. In general, add in the cost of the frame, foundation, and time to cover the cost of cleaning the frames. You may want to sell the frame of comb for so much plus so much for the frame, which would be refunded if returned or deducted if the customer purchases another frame of comb.
- [ ] Scrape the frames to clean up the most soiled areas, primarily to remove the propolis from the frames. Store the frames in the freezer or in a dry, well-ventilated, unheated room until needed.
- [ ] When sold, wrap the frame of comb in plastic wrap and then newspaper. Tell the customer how to cut out the comb and store the sections of honey.

PROCESSING CHUNK COMB HONEY

Chunk honey is comb honey cut into strips and inserted in a wide-mouth jar; liquid honey is then poured into the jar until it is full. Chunk honey is an attractive product and allows the purchaser to enjoy both liquid honey and comb honey. One pound and two and one-half pound wide-mouth jars are available at bee supply houses for packaging chunk comb honey.

- [ ] Place frames of comb honey in the freezer for several days. Remove and dry them completely.
- [ ] Place the frames of comb on an oven rack or wire rack over a tray to catch drips.
- [ ] With a sharp paring knife, cut around the edges of the frame and lift off the frame. Store the frame in a bucket to catch the drips.
- [ ] The size of the strips of comb will depend on the width of the jar's mouth and the depth of the jar. For example, the wide-mouth quart mayonnaise jar takes two strips about 2½ by 5½ inches.

☐ Cut the strips to fit the size of your jar in width and depth. Slide the comb into the jar, and then fill the jar with liquid honey. The combs tend to float as the honey is added. A small scrap of comb on top of the chunk combs will hold the combs down on the bottom of the jar when the lid is screwed on. The product looks more attractive if the combs are not floating off the bottom of the jar. If you are using jars for which you cannot get replacement lids, place a square of waxed paper under the cover of the lid that fits.

☐ Cap the jars and store them in a dry, well-ventilated, unheated area, or store them in the freezer. If you cannot store chunk comb honey (or any of the comb honeys) in a freezer or refrigerator set for freezing, do not process too much of it, for it will tend to granulate. Honey granulated in the comb cannot be reliquefied in the comb, so it will have to be liquefied by melting down the combs. Reliquefication is discussed later in this unit.

CLEANING UP STICKY FRAMES AFTER CUTTING OUT THE COMB

Before you replace the foundation in the frames from which you cut the comb, let the bees clean up the sticky mess so the frames will be easier to handle. Place the frames in a super, and about thirty minutes before dark, set the super or supers on top of the hive for two or three days. Remove the supers of frames and replace the foundation to reuse the supers immediately, or store the supers and replace the foundation at your leisure for use next season.

In the North: If you process the frames of comb in the winter when the bees cannot help in clean-up, store the frames in the supers until the bees become active in the spring, and then return the supers to the hives for clean-up. Remove the supers of frames after several days, replace the foundation, and they are ready for use.

A frame-cleaning tool, available from bee supply houses, is handy for cleaning the wax out of the grooved top and bottom bars. If the frame top has a wedge bar, it must be removed and cleaned of wax. Insert the new foundation and the supers of foundation will be ready to be added to the hive when needed.

GRANULATED COMB HONEY

There is absolutely nothing wrong with granulated honey in the comb or in jars. Many people prefer granulated honey to liquid honey, but unfortunately, most of those people are residents of other countries.

Honeys do granulate differently, depending on the nectar source.

Some will granulate with large crystals and become as hard as rock candy. Others will granulate with fine crystals and can be spread as easily as warm butter.

If your comb honey granulates in large, hard crystals, you may be able to cut it into small squares to use as lunch box treats, or you may find a way to package it to sell as honey candy. If your comb honey granulates in fine, mushy crystals, it can be used almost as easily as liquid comb honey on biscuits, toast, etc. If you cannot use the granulated comb honey, liquefy it by melting it down, and recover the wax.

Liquefying Granulated Comb Honey

Before beginning to liquefy granulated comb honey, take inventory to determine how much liquid honey will result. Use the following guide to decide how large a double boiler unit you need or how many times you will have to make the meltdown to liquefy all your comb honey.

3 section boxes of comb honey	=	1 quart of liquid honey
6 plastic rings of comb honey	=	1 quart of liquid honey
1 shallow frame of comb	=	1 quart of liquid honey
1 medium frame of comb	=	1½ quarts of liquid honey
1 quart of chunk honey	=	1 quart of liquid honey
1 gallon of liquid honey	=	12 pounds of liquid honey

Beeswax becomes liquid at approximately 150° F.

Equipment needed:
- ☐ Large pot (16-quart canning pot or 25-quart tub)
- ☐ Small pot (8- to 10-quart pot or 16-quart canning pot)
- ☐ Wire rack to fit inside large pot
- ☐ Wooden spoon or paddle
- ☐ Candy thermometer
- ☐ Stove (Gas, electric, or wood range)
- ☐ Granulated honey comb or chunk honey
- ☐ Meat fork

Step-by-step directions:
- ☐ Place the wire rack in the bottom of the large pot. Place the smaller pot inside the larger pot on the wire rack. The rack allows the water to circulate.

PROCESSING AND STORING COMB HONEY

- [ ] Place this double boiler arrangement on the stove. Do not turn on the stove. If you are using a wood range and it is heating, place a quart of water in the large pot. This double boiler full of comb honey will be heavy, so it is best to have it on the stove when you fill the inner pot.
- [ ] Cut the granulated honey from the section boxes, plastic rings, or frames. Combs cut into small pieces will melt down more quickly. Cut comb into two-inch squares.
- [ ] When the inner pot is full, add water to the outer pot until the water is two inches from the top. Insert the thermometer in the water. Turn on the stove or stoke up the wood range.
- [ ] Bring the water temperature almost to boiling, and then reduce the heat to hold the water temperature at about 180° F. As the granulated comb honey begins to melt, you may be able to add more pieces of granulated comb honey.
- [ ] Stir the melting mass occasionally with the wooden spoon or paddle to distribute the heat evenly in the mass. *A tip:* Wet the wooden spoon or paddle before dipping it in the mass and the wax will peel off more easily after it has cooled.
- [ ] As the mass becomes warmer, reaching 150° F, the wax will begin to melt and rise to the top and solidify in a thin layer. Continue to heat the mass until the wax on top is completely liquid. Turn off the stove.
- [ ] Let the mixture cool overnight. In the morning, the mixture will be cool and the wax will have solidified into a block on top of the inner pot. Cut around the edge of the inner pot to loosen the wax if it has not shrunk away from the edges completely.
- [ ] With a meat fork, impale the block of wax and lift it out of the inner pot. Let the block drain in a tray, or put it in the sink and wash away the honey.
- [ ] The honey can be poured into jars. It probably has darkened from being heated for an extended period and cannot be sold as natural unheated honey, but it can be sold for cooking honey.

Liquefying Granulated Chunk Honey

Granulated chunk honey in jars must be warmed somewhat before it can be poured from the jar into the double boiler arrangement. The best method is to set the jars of honey in a warm-water bath up to the necks of the jars. Warm the water to 145° F until the honey has liquefied sufficiently to allow it to be poured from the jar, and then pull the granulated comb honey from the jar. Do not raise the water bath above 150° F or the wax will melt in the jar and solidify while it is being removed. Wax in the glass jars is rather hard to remove without submerging the jars in boiling water. Do not immerse empty glass jars or jars of cool honey in hot water, because they may break. Most plastic jars will not stand water temperature much above 180° F before they begin to deform or melt.

Liquefying Granulated Extracted Honey

Granulated honey in jars can be liquefied by using the water-bath method suggested previously. Most authorities do not recommend dry heat in liquefying honey, but if you want to disregard that advice, you can place the jars of granulated honey in the oven with the temperature set at 150° F until the honey has liquefied. During the summer it is possible to liquefy granulated honey by placing the jars on their sides on brick or concrete and letting the sun generate the heat.

UNIT 21

Extracting Honey

TIME TO EXTRACT

In the North: The honey flow in some northern areas may run right up until the first frost and perhaps through the first series of light frosts. As fall approaches, the northern beekeeper must plan ahead to remove the honey supers, extract, clean up the sticky supers, medicate the hives to control AFB/EFB and Nosema Disease (which also adds winter stores in the hive), and perhaps re-queen the hive. Most of these activities require participation by the bees, so they must be done before the temperature goes below and stays below 50° F. If the bees are in cluster and cannot break cluster, they cannot participate.

The less-experienced beekeeper should talk to experienced, successful beekeepers in the area to get a more complete picture of the time available to accomplish the things that must be done after the honey flow and before the bees become inactive for the winter. The best wintering methods can be learned from beekeepers in your area who are successful in wintering their hives.

You may have to store your honey supers for a while after removing them from the hive and extract them during the late fall or winter. You will then have to store the sticky supers through the winter, get them back on the hives for clean-up in the spring when the bees become active again, and then store them until needed during the honey flow. In the northern areas, you do not have to worry about ants and wax moths during the winter as beekeepers do in the southern areas.

In the South: You can extract immediately after the honey flow, but there are still several months of warmer weather following the end of the honey flow in the more southern areas. Unless you need the honey to sell, it may be better to let the bees take care of the honey until the weather begins to cool in the fall. Be aware that the wax moths are with you all year round, and it is better to leave the honey supers on the hive until you are really ready to remove the honey. Extract the honey within two weeks, and then clean up the supers and store them away with Paradichlorobenzene (PDB) moth crystals.

In some southern areas, beekeepers must plan to extract, clean up, and medicate early enough to catch a winter flow from various sources. For example, many pepper trees in Southern California were planted as street trees and provide a fairly heavy nectar flow during the winter months. In the San Francisco Bay Area, many old eucalyptus trees come into bloom and produce a good nectar flow from November through January.

In most southern areas, the bees fly nearly every day during the winter, and they will help in cleaning up the sticky frames, with medicating/feeding, and with re-queening. Fall is a good time to re-queen in the southern areas, because the queen has time to lay and keep the hive strong through the winter.

In both the North and South, the honey may granulate in the comb if you wait too long to extract after the supers are off the hives. Mold may form on the cappings of the comb unless it is stored in a well-ventilated area.

Remember: Do not store comb honey or supers of honey to be extracted with moth crystals, or you will taint the honey, making it unfit for human consumption.

Do not let honey granulate in the extractor, storage tanks, or buckets. Granulated honey is easier to liquefy if it is in smaller containers or jars. If you are renting an extractor, you should have all your supers of honey off the hives before you get the extractor. Plan ahead to remove supers so that the last super comes off the hive on the day you rent the extractor or before.

PREPARING TO EXTRACT

(Step-by-step directions for extracting will follow this discussion.) To extract the honey you must purchase, borrow, build, or rent an extractor. Check with a bee association, other beekeepers, 4-H beekeeping groups, and beekeeping shops about renting or borrowing an extractor. If your beekeeping association does not have a group extractor for rental, you might buy an extractor and make it available for rental to help pay for it. The advantage of owning an extractor is that you can use it at any time. For example, you can extract in the middle of the honey flow if necessary to make three supers do the work of four or to make honey available to your customers if you had sold all your honey earlier.

Most two-frame and four-frame extractors (even some six- and eight-frame extractors) are of the type in which the frames (of any size) stand on end against the sides of a square (hexagonal or octagonal) basket. These extractors are often referred to as tangential extractors and may be hand cranked or motorized. When you use a motorized extractor, plug the motor cord into an on-off switch if the motor does not have a built-in switch, because you will be turning the extractor on and off many times while you extract.

Most larger extractors (from six to seventy-two or more frames) are radial extractors, which are usually motorized and should be plugged into an on-off switch if the motor does not have a built-in switch. Radial extractors support the frames (of any size) in notches around a reel, like the spokes on a wheel. As the reel spins in the extractor, honey is thrown from both sides of the combs. Damage to the frames of combs is minimal, because the strain of spinning is not on the side of the comb but against the top bar of the frame and because the honey is being thrown from both sides of the comb at the same time. Breakage can occur if the combs contain pollen stores or granulated honey. Combs containing considerable pollen or granulated honey should be set aside and extracted as a separate load at a reduced speed. This should be done whether you use a radial or tangential extractor.

The extractor most hobbyists use must be placed on a stand or bench to raise the honey gate above the tallest bucket used to strain the honey. The stand will usually have to be made for the extractor, but if you have extra supers you can make a stand by setting the supers down and nailing a piece of plywood (with small nails) to the supers. This trick also can be used for raising an uncapping tank to a convenient height. (See Figure 17-1.) If you are using a small two-frame extractor, you may find it easier to extract with the extractor sitting on the floor; when it comes time to drain out the honey (usually no more than two gallons), raise it onto a stool or bench.

The smaller the extractor, the more necessary it becomes to snub the extractor securely to the stand or bench if you are going to be extracting quite a few supers. Four-frame and larger extractors—even the radials—can be unevenly loaded and will begin to jump and buck. If the extractor (any size that is not snubbed down) begins to rock or buck, slow the rate of turning until the bucking stops. Turn at the reduced speed for about one minute; then stop the turning and turn the frames of comb over in the extractor. Turn just under the speed at which the extractor begins to buck for three to four minutes. Turn the frames of comb around, and finish extracting the first side. Sometimes getting a little honey out of one side and then extracting the second side causes the bucking to stop completely.

Remember: Cold honey extracts less easily than warm honey, because it is thicker or stiffer; it will require warming or a longer spinning time in the extractor. In any area, if the honey temperature is below 60° F, warm the honey by bringing it into the house for several days, or stack the supers of honey over an empty super containing a lighted electric bulb. Within several days, the lighted bulb under the covered supers of honey will bring the temperature of the honey up enough to make extracting fairly easy.

To uncap the combs of honey you must have an electric uncapping knife; a steam uncapping knife; a pair of "cold" uncapping knives, or a pair of long, thin, butcher-type knives. The "cold" knives and butcher knives are heated in near-boiling water; each knife is used to uncap one side of the frame and then is returned to the hot water. Unless you have a steam generating unit, you will find the electric knife to be cheaper and more convenient. If you are using an electric knife, it should be plugged into an on-off switch, because you will be turning it on and off many times while you extract.

In uncapping the frame, stand it on end and slide the knife down the frame along the top and bottom bar, removing the cappings as the knife moves down. Always slide the knife along the top and bottom bars, regardless of how far the comb extends on the frame, because you want the combs drawn out evenly on both sides of the frame when the bees use it again. If you leave extended comb on one side of a frame, the frame next to it will have to be shallow to provide for bee space, and you will always have to use the cappings scratcher to remove the cappings on the shallow side. Any cappings that are too low to be removed by the knife must be removed by tearing the cappings open or lifting them off with a cappings scratcher or a fork. Some beekeepers with only one or two hives use the cappings scratcher rather than a knife to remove the cappings. This is considerably slower but, if time is of no consequence, does the job very nicely.

The honey must be strained. You can use a sieve, cheesecloth, clean nylon panty hose, or a settling tank. Sieves and cheesecloth clog after a short time, so you should have several on hand. The settling tank may be a bit expensive, but one can be made from a fifteen- to twenty-gallon plastic tank by inserting a small plastic gate, obtained from a bee supply house. The settling tank will be discussed more fully later in this unit.

The best way to strain honey is to use panty hose. Drop the legs in a bucket, and stretch the waistband over the top of the bucket. Extract and fill the bucket; then release the waistband and tie it with a stout cord. Toss the cord over a rafter, and lift the panty hose out of the bucket. Raise the panty hose about one-third of the way out of the bucket and let drain for about two minutes; then raise the hose another third of the way and let drain again. Finally, when the toes of the panty hose are a few inches above the rim of the bucket, let them drain for about ten minutes, then slide the bucket aside while you slip another bucket under the panty hose. Lower the panty hose into the bucket, replace the waistband around the bucket, and you are ready to strain another bucketful.

As the honey drains down the inside of the panty hose, all the wax bits end up at the toe and the strainer never clogs. If you find the panty hose stretch too long, next time you can tie off each leg in the middle and cut off the excess, which will keep the stainer from stretching out over six feet long as it is being raised up to the rafter.

The panty hose will remove larger crystals of granulated honey if you have had some granulation in the comb and the granules are thrown out in the extracting. After the extracting is done, let the panty hose drain for several days, and turn the hose inside out to salvage the wax scraps for melting down with the cappings. (If you have not discovered this already, panty hose can be used for straining many things, such as paints and other liquids which have chunks or granules that need to be removed.)

Make the extracting area as bee-tight as possible. This may mean chinking the openings over and under the garage door with sheets or other rags. You can extract in the house or kitchen, but if you do you should spread out newspapers or plastic sheeting, because there will be drips of honey and wax bits. When the author extracts he keeps a clean pair of shoes or slippers handy so he can slip into them if he leaves the extracting area to go into the house or (if extracting in the kitchen) into another room. If you have no place in which to extract except out in the open, you must do your extracting after dark. Once the bees have settled in for the night you can have a light on, unless you are extracting very near the hives. If you cannot finish extracting out in the open in one night, bring all your equipment inside or wrap the extractor in a sheet and bring everything else in so the bees will not be able to get at it the next day. Once the author had to extract in the open. He put all the utensils, cappings, etc., in the trunk of his car and wrapped the extracter in a beach towel so the bees could not get in during the day. After dark he took everything out, unwrapped the extractor, and began extracting again.

To gauge roughly how much honey you will have after extracting, use this guide.

One deep super of comb	=	3 gallons of liquid honey
One medium super of comb	=	2 gallons of liquid honey
One shallow super of comb	=	1½ gallons of liquid honey

EXTRACTING THE HONEY

Two extraction methods are described in this section: Method 1 for the small operation and Method 2 for the modest operation. As your operation becomes larger, you will find ways to make extracting easier and faster.

Method 1: Extracting for the Small Operation

Equipment needed:
- [ ] Extractor—2-frame, 4-frame, or larger
- [ ] Uncapping knife
- [ ] Cappings scratcher or fork
- [ ] Cake pan for uncapping—about 2" deep
- [ ] Collander for draining the cappings
- [ ] Several sieves or a pair of panty hose
- [ ] Pan of water and washcloth
- [ ] Buckets and jars
- [ ] Hive tool
- [ ] Bowl scraper

Step-by-step directions:
- [ ] Before removing any frames from the super you are going to extract, use the hive tool to remove the burr comb from the top bars of all frames in the super. Save the scrap wax in a can, bag, or box.
- [ ] Plug in the electric knife, if you have one. Remove a frame of comb from the super, and stand it on end in the cake pan with the end of the frame top bar over the edge of the cake pan.
- [ ] Place the knife at the top of the frame across the sides of the top and bottom bars. Tilt the knife at about 45°, and slice back and forth as you press or push the knife slowly down to the lower end of the frame. As you slice down the frame, the cappings will fall away from the knife into the cake pan. Some low areas on the comb will not have the cappings removed. Do not worry about those areas at this time. (See Figure 21-1.)
- [ ] Turn the frame over in the cake pan and repeat the previous steps, removing the cappings from the opposite side of the frame. Do not worry about any low spots at this time. Place the uncapped frame, with any capped low spots, in the extractor, which will catch any drips from the uncapped cells.
- [ ] Remove another frame of comb from the super, and following the previous steps, uncap both sides of the frame. Continue uncapping

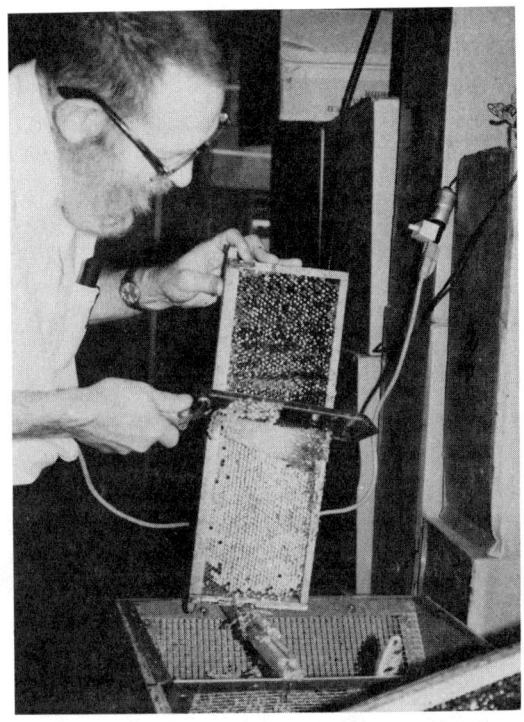

Figure 21-1, Uncapping a Comb of Honey. Slicing back and forth down the frame with the blade on the side of the top bar and bottom bar. Notice that the electric knife is attached to an on-off switch. In this photo, the cappings fall into a basket suspended on the capping tank. This tank will hold about sixteen frames of honey; when uncapped, it will catch the dripping honey. (See Figure 21-4.)

frames until you have done enough to fill the extractor. Turn off the electric knife or it will continue to heat, charring honey on the blade and reducing its efficiency. When the electric knife is turned off, cool the blade by cutting through the cappings in the cake pan in several directions. This not only cools the blade to stop any charring of honey but also makes the cappings drain more readily when they are dumped from the cake pan into the collander.

☐ Remove one of the frames from the extractor, and stand it in the cake pan as you did when you uncapped it. With the cappings scratcher, tear and break open all the cappings of cells too low to be uncapped by the knife, or slip the cappings scratcher just under the capped cells, push about one inch, and lift. The push and lift will remove the caps in one piece. You may notice as you are uncapping the low spots that the areas where the electric knife cut away the cappings it left a very thin sheen of wax, covering the cells that were uncapped. Generally the honey will be thrown through this thin sheen, but the author usually drags the tines of the cappings scratcher gently over those areas to break the covering.

Continue to tear or push and lift until all the cappings have been removed from both sides of the frame, and then return it to the extractor. (See Figure 21-2.) Repeat these steps until all the frames in the extractor have all cell caps torn open or removed.

In scratching the cappings with the cappings scratcher or fork, draw the fork across the cappings in one direction and then in another and another until all the cappings have been broken or torn. Use enough pressure to tear the cappings. Full cells of honey are reasonably strong, and any broken edges will be repaired by the bees before being refilled next season. In fact, the bees will tend to draw these cells out a bit farther than they were before. Eventually you will not have to scratch or lift many low spots.

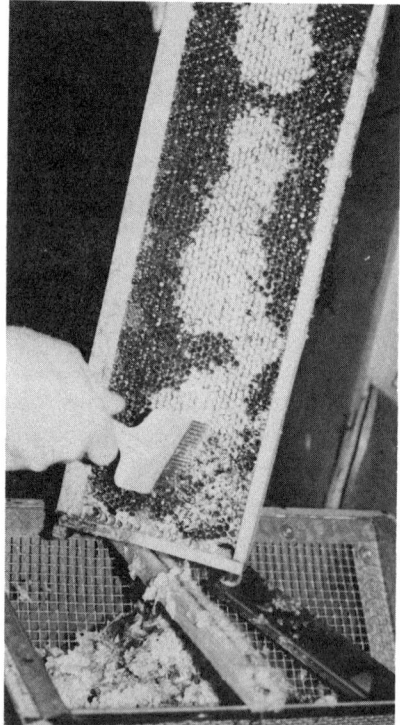 

Figure 21-2, Using the Cappings Scratcher. Left: The cappings that were too low to be removed by the knife are being torn loose by the sharp tines dragging back and forth over the cappings. Right: The tines are slipped under the cappings and pushed forward about one inch. Raising the cappings scratcher lifts off the cell cappings in one piece.

☐ *Now you can begin to extract.*

A word of caution is necessary here. Unless you are using a radial extractor, it is important to remember that the first side of the combs being extracted must be spun more slowly and, therefore, longer. The reason is this: Centrifugal force throws the honey from the cells. The cells of honey on the back side of the frame are weighted and will tend to be thrown through the midrib of the comb. If spun too fast on the first side, the comb can be torn apart or cracked along its middle by the weight of the honey on the back side of the frame.

Spin the combs in the extractor more slowly over a longer period of time and the weight on the back side of the comb will not become great enough to break the comb. When the first side has been emptied and the comb turned over, there will be no weight on the back side of the comb, and the comb can be spun at a higher rate of speed on the second side of the comb.

If honey has granulated or pollen has been stored in the comb, you must extract more slowly and longer on both sides of the frame so you will not damage the comb. Always try to get balanced loads in the extractor. Do not place a full frame opposite a half-empty frame; try to get two half-empty frames opposite one another. Unbalanced loads will cause the extractor to jump and shuffle around the room.

☐ Finally you can begin to turn the crank or turn on the motor, and the basket holding the uncapped frames will begin to spin. Spin the basket (not necessarily the crank) at about 100 RPM. Some small extractors have a one-to-one cranking; that is, one turn of the crank is one turn of the basket. Most extractors with a side crank and gears have a three-to-one ratio, which means one turn of the crank equals three turns of the basket; so in this case, you would turn the crank at 33 RPM to have the basket turn at 100 RPM. If you are using a motorized tangential extractor, set the speed so the basket will turn no faster than 100 RPM on the first side. Slower might be better, and you can be uncapping and scratching two or four more frames while the motor is turning the basket. If you have no uncapping tank, you can stand two or three uncapped frames of comb in a five-gallon bucket to catch the drips. The extractor can spin slowly for five or ten minutes without hurting anything but the utility bill.

☐ Turn the basket for two to four minutes on the first side. Go for three minutes on the first set and work up or down from there, depending on how warm or how thick or sticky the honey is. Some honeys are stickier or stringier than others and take longer to remove from the cells in extracting, even though the honeys may be at the same temperature. In the San Jose, California, area it seems the season's first super of honey is stringy when extracted, whereas the later supers seem to extract rather smoothly.

☐ After three minutes, stop the spinning. Lift the combs and turn them around. The full side of the comb should be facing the basket and the tub of the extractor. Begin to spin again. Now you may spin the basket as fast as you can, just under the speed at which the extractor begins to rock or wobble because it is unbalanced. Spin the basket for about one minute. Stop the spinning.

☐ Remove the frames from the extractor, and return them to the super.

☐ Repeat the steps for uncapping, scratching low spots, and extracting until all frames in all supers have been extracted.

Do these things as you go along:

When the cappings begin to pile up in the cake pan, chop them once again with the knife and dump them to drain into a collander that is placed on a bucket.

Drain the honey from the extractor. Depending on its size, the extractor may hold the honey from six to sixteen frames in the tank. Drain the extractor when required or as often as you like. (See Figure 21-3.)

Important note: When you drain the extractor, you should not do anything else—you may forget the honey is draining if you get involved in some other activity. Quarts or gallons of honey may spill onto the floor; besides wasting all that honey, you will have a miserable mess to clean up. If all your bees are healthy, you might be able to feed them the spilled honey in the spring. The author now has a more-or-less failsafe system of setting his buckets in a tub, so he's OK until the bucket and the tub both fill and begin to overflow. You can buy a battery-operated alarm that can alert you to the fact that your bucket is full and will soon run over. When set up, this system has a ball-type float that is supported about two inches inside and below the top of the bucket. When the ball is lifted by the honey as the bucket fills, it turns on the alarm bell. You should then stop whatever you are doing and close the honey gate on the extractor.

Strain your honey as you drain it from the extractor, using the sieves or the panty hose as mentioned earlier in this unit. If you use a settling tank, you can take the honey drained from the extractor without straining and pour it directly into the tank. See Method 2 for more discussion of the settling tank and how it is used most efficiently. Use a pan of water and washcloth to clean up sticky hands and the handle of the knife and cappings scratcher once in a while.

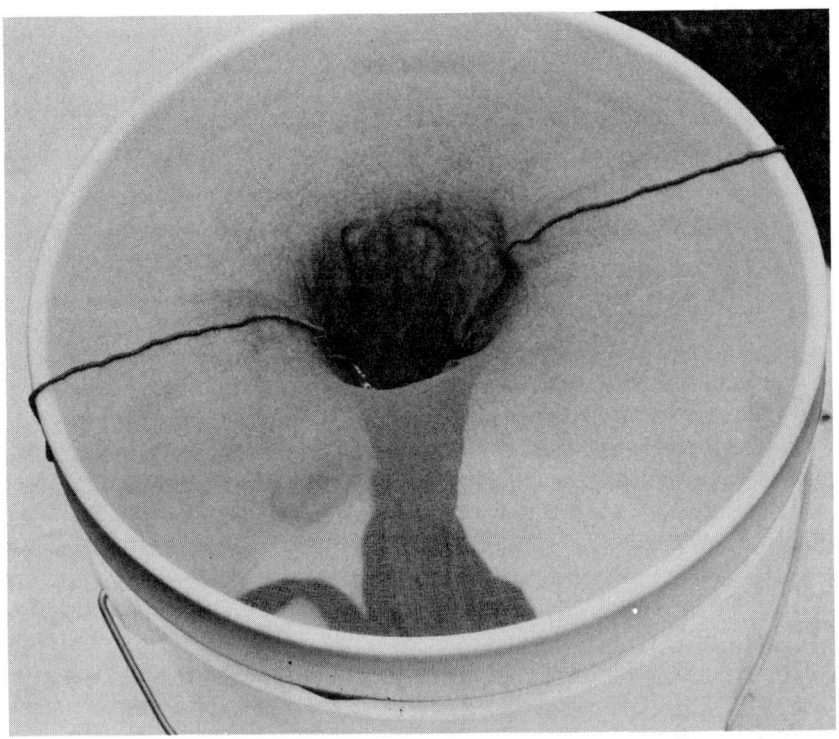

Figure 21-3, The Panty Hose Strainer. Above: The panty hose are on a bucket ready to be filled with honey from the extractor. Notice that the legs of the panty hose are down inside the bucket, and the waistband is stretched around the bucket. Right: The waistband has been released from the bucket and fastened with a stout cord that is raised over a rafter or by a pulley arrangement. The panty hose legs are hanging just out of the bucket as the honey drains. The hose can be used to strain many buckets of honey before they wear out.

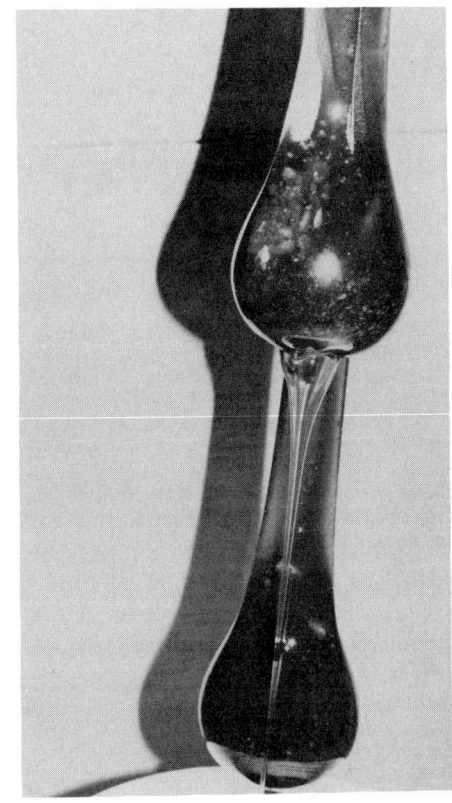

Method 2: Extracting for the Modest Operation

Equipment needed:
- ☐ Extractor
- ☐ Uncapping knife
- ☐ Cappings scratcher
- ☐ Uncapping tank and cappings basket*
- ☐ Settling tank**
- ☐ Pan of water and washcloth
- ☐ Buckets and jars
- ☐ Sieves or panty hose
- ☐ Hive tool
- ☐ Bowl scraper

Step-by-step directions:
- ☐ Before removing any frames from the super you are going to extract, use the hive tool to remove the burr comb from the frame tops of the super before starting to uncap the frames. Toss the burr comb in the capping basket, or save it in a box, bag, or can. Burr comb on the sides and bottom bars of the frames is best removed when you uncap; otherwise, it can cause a messy drip problem.
- ☐ Plug in the electric knife. (It is more convenient if the electric knife is plugged into an on-off switch.) Remove a frame for uncapping.
- ☐ With the cappings basket at a convenient height, set the frame to be uncapped on the nail in the middle of the wooden strip, and begin. Place the edge of the knife at the top of the frame, and begin to slice down the sides of the top and bottom bars. Hold the knife against the frame at an angle of about 45°, and use a back and forth

---

*For the moderate or larger operation, buy an uncapping tank and a cappings basket. The tank holds the cappings basket and about fourteen to sixteen frames of comb. The frames can be uncapped and scratched into the cappings basket, then placed in the tank, where they will drain until placed in the extractor. (See Figure 21-4.)

For the modest operation, you can make a cappings basket from a tub. Make a frame that will stand or be held by the sides of the tub about four inches from the bottom of the tub. Cover the frame with ¼-inch hardware screen. The screen will catch and hold the cappings, and they can drain through into the tub. Across the top of the tub place a 1-by-¾-inch strip of wood long enough to reach across the tub. Cut grooves on each end so the strip will fit down over the rim of the tub and will not slide around. Drive a nail up through the middle of the strip to hold the frame on the nail point for easier uncapping. Seriously consider having a hole cut near the bottom of the tub to accept one of the plastic honey gates so you can drain the honey from the tub without having to manually remove the cappings to drain the honey.

**The advantage of the settling tank is that it eliminates straining the honey and permits bottling jars of honey directly from the settling tank.

As extracting proceeds, the honey is poured into the settling tank. It is not necessary to strain the honey, because it is allowed to set for two days when the tank is full, and

Figure 21-4, An Extracting Setup for a Moderate-sized Operation. Shown is the uncapping tank holding a number of medium frames of honey ready to be uncapped. The uncapping tank holds the honey draining from the cappings basket (see Figure 21-1) and the dripping honey from the uncapped honey frames. Honey can be drained away by the honey gate on the end of the tank. The extractor is near the uncapping tank for loading. Notice that empty supers are used to raise the tank and extractor to a convenient height.

movement as you slice to the bottom of the frame. As you cut, the cappings will fold back and drop into the cappings basket. You may slice from the bottom of the frame to the top, but notice how the cappings and honey slide across the blade. This action absorbs heat from the blade of the knife and takes more time for the knife to cut smoothly. In slicing down the frame, only the edge of the blade

---

all the wax will rise to the top. Jars may be filled with clear, clean honey from the honey gate near the bottom of the settling tank. Fill jars or buckets from the tank until the honey level is one inch above the honey gate. If you fill the tank several times from your extracting, just continue to add unstrained honey to the tank with the wax and honey already in the tank. When the tank is full again, let it set for another two days, and all the wax will rise to the top, so again you can draw clear, clean honey into jars and buckets. When the extracting is done and the last of the clear, clean honey has been drawn off to within one inch of the honey gate, then it is time to drain off the last few gallons of wax and honey in the tank into the panty hose and strain the wax from the honey as suggested earlier in this unit.

For the moderate or larger operation buy stainless tanks that will hold 300 or 600 pounds of honey. The modest operator can make a settling tank by obtaining a sturdy (heavy-duty) plastic garbage can of twenty- to thirty-gallon capacity. Buy a plastic honey gate and attach it by cutting the proper opening about one inch above the bottom of the garbage can. The honey gate has a rubber ring and a plastic nut, which secures the gate and seals against leakage. (See Figure 21-5.)

Figure 21-5, Settling Tanks. Two plastic containers into which holes were drilled or cut and plastic gates were installed. The larger container can be used to fill jars after the wax particles are given twenty-four to forty-eight hours to rise to the top. The five gallon bucket can be used to fill jars after the panty hose are used to strain out the wax bits.

comes in contact with the cappings and honey, so less heat is lost as you cut.

☐ Make only one pass on each side of the frame with the uncapping knife. In some spots, the cappings may be too low to be removed by the knife as it slices down the sides of the top and bottom bars. Place the frame in the extractor to catch the dripping honey.

☐ Repeat the previous steps until enough frames have been uncapped to fill the extractor or the uncapping tank.

☐ Turn off the electric knife. Cut through the cappings in the tub to cool the blade. This slicing of the cappings as you go along will help them to drain more completely.

☐ Remove a frame from the extractor or uncapping tank, and set it on the nail in the strip across the tub. With the cappings scratcher, tear or lift off the cappings in the low spots on each side of the frame. Drag the scratcher along the low spot in one direction, then in another and another, to make sure the cappings are broken so the honey can be thrown from the cell during extraction. In lifting off the cappings with the scratcher, push the tines just under the cappings and push forward about one inch; then lift the scratcher.

The cappings over the tines will lift off in one piece. Repeat pushing and lifting until all the low spots have had the cappings removed. Replace the frame in the extractor.

- [ ] Repeat the previous step until all the frames in the extractor have had the cappings in the low spots scratched or lifted off. Use the pan of water and the washcloth to wipe off your sticky hands or the handle of the uncapping knife or cappings scratcher.
- [ ] Spin the basket in the extractor. The speed on the first side should be about 100 RPM of the basket (not necessarily the crank). Some small extractors have a one-to-one cranking; that is, one turn of the crank equals one turn of the basket. Side-crank extractors usually have a three-to-one turn ratio; that is, one turn of the crank equals three turns of the basket. Using motorized extractors, you must guess at a rotation of about 100 RPM for the first side of the frames. A slower speed would not hurt, because the motor is doing the work and it could run longer, if necessary. Spin the frames for three or four minutes. If the extractor is motorized, you can spin the basket slower while you are uncapping and scratching other frames of comb. If you have no uncapping tank, you can stand three uncapped frames in a five-gallon bucket to catch the drips.
- [ ] Turn the frames over in the extractor, with the full side of the frame facing the extractor tub. Spin the basket as rapidly as possible just under the speed at which the extractor begins to rock or jump. Spin the frames for one minute on this side.

    If you are using a radial extractor, you do not have to worry about speed, because the frames are supported in a reel like the spokes in a wheel. The top bar takes all the stress in the extracting, and the honey is extracted from both sides of the frame at the same time. Once the radial extractor has picked up speed, you can set it to go as fast as possible without jumping, for about ten to twelve minutes.
- [ ] Remove the frames from the extractor, and replace them in the super.
- [ ] Following the previous steps, uncap, scratch, and extract—load after load—until all the frames in all the supers have been extracted.
- [ ] **Remember:** You will have to drain the honey from the extractor tank at intervals, because most extractors cannot hold the honey from more than six to sixteen frames. The honey must be drained before the top bar ends begin to drag in the honey, which would make the extractor hard to turn or the motor unit spin without turning the extractor basket.
- [ ] Strain the honey. You can use the panty hose method as mentioned earlier in this unit, or if you have a settling tank, proceed as follows.

As you extract, drain the honey without straining into buckets, and pour it into the settling tank. When the settling tank is full, let the honey set for two days while the wax rises to the top. If the weather is very cold, you may have to use the panty hose strainer, for the honey may be too cold and stiff for the tiny wax particles to rise to the top in the settling tank.

- [ ] *After two days:* Fill your jars and buckets directly from the honey gate of the settling tank. The honey will be clean and clear all the way down to about one inch above the top of the honey gate. If you still have more frames of honey to extract, proceed with the extracting and fill up the settling tank again, wait two days, and bottle the honey. When all your honey is extracted and bottled from the settling tank down to about one inch above the gate, use the panty hose strainer for the last few gallons of wax and honey in the settling tank.
- [ ] Tilt the extractor up to drain it as completely as possible, remove the extractor basket, and scrape the remaining honey from the sides of the extractor with a rubber bowl scraper.
- [ ] After the settling tank has been tilted and drained into the panty hose strainer, take the rubber bowl scraper and scrape the honey from the sides of the tank.
- [ ] Now the cappings basket must be drained and the cappings moved into buckets or a solar melter or melted down in a double boiler arrangement. In general, cappings are about 20 percent wax and 80 percent honey by weight. Get the cappings out of the cappings basket, and drain the honey out of the tank or tub so it can be strained through the panty hose and you will have the bulk of the honey available to bottle. If you do not have a honey gate in your tub, the best way to transfer the cappings is to roll up your sleeves and physically remove the cappings by hand into buckets, then drain off the honey for straining. If you want to let the cappings drain for a week or two before melting them down, you can pour the cappings back in the cappings basket after draining out the honey. Let the panty hose strainer drain into the cappings basket or into a can or pan for several days before dumping that wax into the cappings basket or melting it down in the solar melter or double boiler.

CLEANING THE EXTRACTING EQUIPMENT

Probably the easiest way to clean the extractor tank, extractor basket, and all the other items is to load them up and take them to a coin-operated car wash. The hot water or steam, if available, will clean things up in a hurry. Some beekeepers attach a hose to the hot water heater and use the hot water pressure through a hose nozzle to clean up the extracting equipment. Some beekeepers take the extractor and basket into the shower or bathtub to clean them up. Otherwise, warm or hot soapy water can be used to wash out all the extracting equipment except the electric knife and the extractor motor. After washing with soapy water, rinse the equipment with water from the hose. After the extractor has drained for about ten minutes, dry the equipment with a towel.

Another cleaning method that most beekeepers immediately think of is to let the bees clean up the sticky equipment before washing it. This is not recommended, because the foulbrood diseases can be spread this way. If all your hives are disease-free or you just do not heed good advice, the following steps will keep you from having fighting, stinging bees in your yard or neighborhood.

☐ Drain and then use the bowl scraper to scrape the extractor tub, buckets, pans, etc., as clean of honey as possible; otherwise, hundreds of bees may drown in the honey that will collect overnight in the lowest part of the containers.

☐ *This is the important step.* After dark, and only after dark, set the pieces of equipment, extractor basket, extractor tub, buckets, pans, sieves, etc., around the yard at least fifteen feet from the hives.

☐ Early in the morning, the bees will find the sticky equipment and call out the reserves. Usually within three or four hours the equipment will be cleaned up and ready for washing. Do not make the mistake of placing the sticky equipment out for clean-up in the middle of the day. Bees from all over will find the equipment, and more fighting than cleaning will go on. Bees are not happy when they are fighting and may sting anyone in the area.

UNIT 22

Cleaning and Storing Supers of Extracted Comb

There is some controversy about whether supers of extracted comb stored without clean-up cause granulation to occur more rapidly in the extracted honey in following years. There is no doubt that such frames are a sticky mess to handle, and if not cleaned the next spring, they will attract ants and wax moths during the warmer weather. If you can get the bees to clean up the sticky supers after you extract in the fall (before you store the supers), you will not have to handle the supers again until they are needed during the honey flow the next year.

In the North: Only in the North will there be a problem of whether to store the supers of extracted comb sticky or clean. It may be because of either necessity or procrastination that the honey is extracted at a time when the bees are inactive for the winter, and there is *no* way the comb can be cleaned without the help of the bees.

If you can, plan to remove the honey supers as soon as the honey flow ends; extract, give the supers back for a few days for clean-up, remove the cleaned supers, begin the medication schedule, and store the clean honey supers. Another activity you might consider is this: Remove and extract all but one super from each hive shortly before the honey flow ends. Let the bees clean up those supers, and then store them. This way you will have only one sticky super of comb from each hive to store through the winter, and those will be the first honey supers you will put on the hive in the spring.

In the South: In areas where the bees fly almost every day, the urgency to get the honey supers off and extracted before the bees become inactive is not a problem. The urgency in the South is to get the supers extracted, cleaned up, and stored before wax moths begin to infest the combs. In some areas of the South, some flora blooms in abundance in the winter and can produce a fair flow of honey. If you are in one of those areas, you will want to extract, clean up, and store the supers.

Then perform the medications before these flora begin to bloom so that medication will not be stored in the super you put on the hive to catch the winter flow. (It was mentioned earlier that in certain areas of California, eucalyptus and pepper trees produce rather heavy flows of nectar during the winter months.)

If there is any question whether you have winter flow in your area, add a honey super the first winter. When you check in the spring and you find there are stores worth gathering, super for the winter flow each winter after you have medicated the hives.

CLEANING UP SUPERS OF STICKY EXTRACTED COMB

Equipment needed:
- [ ] Hive tool
- [ ] Smoker or hive bomb
- [ ] Bee escape
- [ ] Queen excluder, if one is not on the hive
- [ ] Inner cover, if one is not on the hive

Step-by-step directions:
- [ ] Wait until *one-half hour before dark.*
- [x] **Dress properly.**
- [x] **Light the smoker** or have hive bomb handy.
- [x] **Enter the hive.** Place the inverted hive top on the ground nearby.
- [ ] Place the queen excluder over the brood chamber, if it is not there already. Set the inner cover over the queen excluder.
- [ ] Bring the supers of extracted comb from the extracting area, and set them on top of the hive one by one. They will be sitting on top of the inner cover. It is a good practice to put the supers and combs from each hive back on that hive, just in case you have not detected a diseased hive in your apiary.

There are two reasons for setting the extracted supers on the inner cover. First, the bees will bring the honey from the sticky comb down into the brood chamber more readily if they have the restriction of coming and going through the oval hole. Maybe they think they are robbing the super. Second, after several days, you can just tilt the supers up a bit and slip in the bee escape to clear the empty supers of bees over the next twenty-four to forty-eight hours.

The reason for having the queen excluder on the brood chamber is to make sure the queen will not run up into the combs of the

honey supers and begin to lay. If you leave the supers on the hive for very long, the bees may begin to put honey in them; when those supers are stored for the winter, this honey could taint next year's honey.
- [ ] *Wait two to four days—no more!*
- [x] **Dress properly.**
- [x] **Light the smoker** or have hive bomb handy. *Do not* enter the hive or remove the hive top.
- [ ] Smoke between the inner cover and the honey super that is on the inner cover. Wait one minute. While waiting, pull the stack of honey supers toward you about one inch. Tilt the supers up at the inner cover about the height of the hive tool, and prop them up with the hive tool while you reach in and drop the bee escape in the oval hole of the inner cover. Lower the stack of supers, and push them back into place on the inner cover.
- [ ] *Wait two days.*
- [x] **Dress properly.**
- [x] **Light the smoker** or have hive bomb handy.
- [x] **Enter the hive.** The bees should have left the honey supers empty and clean.
- [ ] Remove the supers one by one, and carry them to the garage or other bee-free area. Return to the hive and smoke under the inner cover. Wait one minute.
- [ ] Remove the bee escape from the inner cover, and then remove the inner cover and the queen excluder. Replace the inner cover on the hive (rim side up). Store the queen excluder until needed for a winter flow, if you have one in the South, or until spring, to get it out of the hive where the constant condensation during winter will cause it to rust more quickly.

STORING CLEANED-UP SUPERS OF EXTRACTED COMB

When the clean supers of comb have been removed from the hive, proceed to store the combs under Paradichlorobenzene (PDB) moth crystals. Do not buy or use the napthalene moth crystals, for they just do not seem to work against the wax moth.

In the North: Even if you store your supers sticky, have the bees clean them up in the spring, and then store all but the first honey super until they are needed. In this way the wax moth will not get to your combs. If you are going to raise brood above the queen excluder into the first honey super, or if you have been doing it, remember to keep the honey super with the dark combs for your first honey super in the spring, and store the others.

In the South: Be sure to store the supers of comb with PDB moth crystals anytime they are not on the hive, at any season.

Two methods of storing the supers of comb are described here. Method 1 is for up to about thirty supers. Method 2 is for up to about 100 supers.

Method 1: Bagging Clean Honey Supers

Equipment needed:
- [ ] Paradichlorobenzene (PDB) moth crystals
- [ ] Large plastic garbage bags
- [ ] Tablespoon
- [ ] Supers of clean, empty comb to be stored

Step-by-step directions:
- [ ] Stand a super of comb on end. Open and slip a large plastic bag down over the super. Turn the super over, end over end. This is easier than trying to slip the super into the bag.
- [ ] Drop two or three tablespoons of PDB in the bag with the super of comb. Fasten the bag with a bag tie or tape. (See Figure 22-1.)
- [ ] Take care not to tear the bag as you insert the super or as you stack it in an undisturbed area of a shed or garage. If you can keep the adult moth out of the bag, it will be safe for months or years. The moth crystals will kill the wax moth larvae when the eggs hatch. Unless an adult moth can get into the bag through a tear, your supers will be safe for a long time.
- [ ] Continue bagging and storing the supers until all are bagged and stored. If you are going to raise frames of brood above the queen excluder in future years, mark the bag containing the super with dark comb to be used as the first honey super in the spring.

Method 2: Stacking Clean Honey Supers

Equipment needed:
- [ ] Paradichlorobenzene (PDB) moth crystals
- [ ] Tablespoon
- [ ] Bundle of newspapers
- [ ] Sheets of paper on which to place the PDB

234 KEEPING BEES

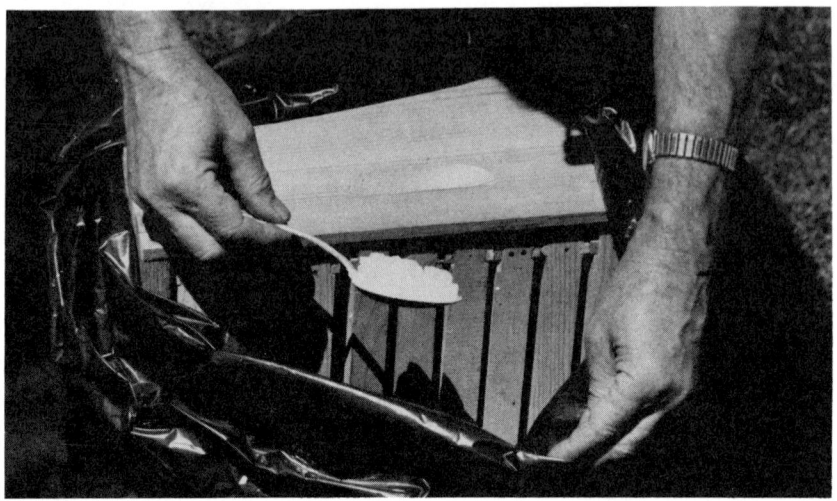

Figure 22-1, Storing Clean Honey Supers. After the bees have cleaned the honey residue from the extracted combs, the super with frames is placed in a large plastic garbage bag. Two tablespoons of PDB moth crystals are dropped into the bag. The bag is then tied off securely and stacked carefully so as not to tear.

Step-by-step directions:
- [ ] Place about three layers of newspaper on the floor of the garage or shed where the supers will be stored.
- [ ] Place a super of comb on the papered area, and put a small sheet of paper on the frame top bars. Place two tablespoons of PDB on the small sheet of paper. (See Figure 22-2.)
- [ ] Place a second super of comb on the first super. Add two tablespoons of PDB on a small sheet of paper on the frame tops.
- [ ] Place three layers of newspaper over the two stacked supers. Place the next super of comb on the newspaper over the two stacked supers. Repeat the steps above, continuing to stack supers, each with PDB on a small sheet of paper and each pair of supers divided by three layers of newspaper. When the stack is as high as you can reach, start a new stack. Place three layers of newspaper on top of each stack, and place a queen excluder on top to hold the paper down and to keep mice and rats from the comb.
- [ ] *Wait three weeks.*
- [ ] Restack the supers. As you restack the supers, place two tablespoons of PDB on the small sheets of paper. The reason for the restacking is that the supers are not tight (as were the plastic bags),

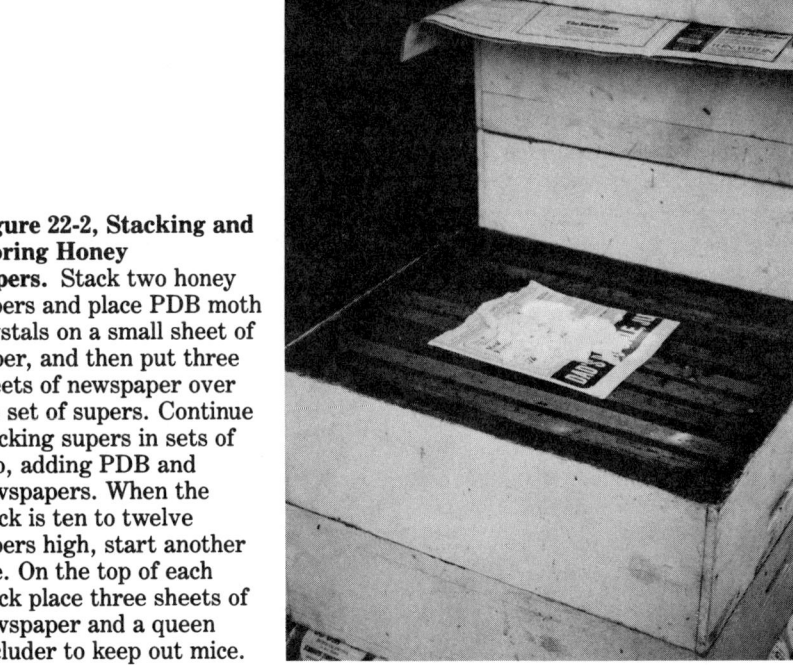

Figure 22-2, Stacking and Storing Honey Supers. Stack two honey supers and place PDB moth crystals on a small sheet of paper, and then put three sheets of newspaper over the set of supers. Continue stacking supers in sets of two, adding PDB and newspapers. When the stack is ten to twelve supers high, start another one. On the top of each stack place three sheets of newspaper and a queen excluder to keep out mice.

and the air leaks in and out of the cracks. In order to insure a good kill of all larvae in the supers, it is necessary to add more crystals to keep up the concentration of gas in the supers over a longer time. If the supers fit snugly together, the adult wax moth cannot get in. If you have supers with nicks and warps, you should store those with moth crystals in plastic bags. If you plan to store supers for several seasons, do not use this method, but store them as directed in Method 1.

HANDLING STICKY SUPERS OF EXTRACTED COMB

In the North: If you extract late, either by design or by accident, the bees will be inactive and the temperatures will soon be below 45° F. Stack the supers where they will not be disturbed or knocked over during the winter. Cover the stacks of supers with a queen excluder or plywood to keep out mice and rats. In the spring, after the bees become active again, clean up the supers as directed earlier in this unit. When the supers are cleaned and removed from the hive, store all but one super per hive with PDB crystals. The one super per hive will be the first honey super. If you will be raising brood above the queen excluder into the first honey super, be sure the super you save for the first honey super is the one with the dark combs.

STORING A HIVE THAT DIED OUT

In the North: If you are going to restock the dead hive, move it to a bee-free area, or close it off completely. You will not need to store the brood chambers with PDB unless you are going to store them over a season before restocking.

In the South: Anytime a hive dies out, immediately get each brood super into a plastic bag with PDB moth crystals as directed earlier to store clean honey supers. Store the brood supers in moth crystals if they can be reused again for another package or swarm. If the combs are badly damaged by wax moths, put in new foundation. Always check for AFB/EFB symptoms. Air the supers for several days before restocking a hive on stored brood comb.

PART **VI**

Preparing for Winter

Winter is one of the most critical periods in the life of the hive. The next three units cover the activities and manipulations necessary to prepare the hive for winter.

- **Unit 23** FALL MEDICATION AND FEEDING OF THE HIVE
- **Unit 24** UNITING WEAK HIVES
- **Unit 25** WINTERING THE HIVES

UNIT 23

Fall Medication and Feeding of the Hive

FALL MEDICATING AND FEEDING

The bees must be active in order to take fall medication or feed. Medication for the control of the foulbrood diseases is always important. Feeding of the hives is more important in some areas than in others. If you are going to feed your hives, medicate the sugar solution to control Nosema Disease.

In the North: As soon as the extracted honey supers of comb have been cleaned up for storing, begin the AFB/EFB medication. If you are not going to clean up the supers, begin to medicate as soon as all honey supers have been removed.

It is important to depend on medication to aid in the control of AFB, because the population decreases in the fall to a winter cluster of about 25,000 bees. This may not be enough bees to quickly clean out the larvae that die of the disease. The medication will keep most of the larvae from dying at the end of the season and act to control the disease. The larger population of bees, before and during the honey flow, can more easily keep the disease under control; by removing dead larvae more quickly, fewer AFB spores will be spread in the hive.

If re-queening was in your plans for the fall, you should have ordered your queen earlier and have her in hand just before or very soon after the flow ends. A young queen will lay longer into the fall and will lay more heavily during that time (if nectar/honey and pollen are available). If you re-queened in the spring, it is very likely that your spring queen will be a good layer into the fall, and you should have a nice population of young bees if nectar/honey and pollen are available. If you are not re-queening in the fall, it is important to inspect the brood chamber at the time you begin medicating to make sure that the queen is there and that she is laying. Do not expect so large an area of brood as was found in the spring or during the summer, because nectar in the fall will arrive in smaller quantities, and the bees will not feed the queen so heavily.

One advantage of feeding the hive in the fall (or medicating to control Nosema Disease) is that it keeps the queen actively laying longer into the fall. When nectar or sugar solution is coming in and being stored, the bees feed the queen more heavily, and she continues to lay. Remember that young bees are an important wintering requirement. When you feed the bees at the same time you medicate with the AFB/EFB Terramycin®-powdered sugar/Drivert® sugar mix, you spread the AFB medication over a larger area of the brood, which is an effective control of AFB.

Beekeepers in the far North (and in higher elevations in any area) may find that winter will sometimes set in by late September or early October, so make plans to move as quickly as possible; do not get caught with your hives unprepared for winter. If you miscalculate on the time available for medicating or feeding, you will have to take the consequences, for there is nothing you can do about the weather—except complain about it.

In the South: In the far South, where the bees fly almost every day, there is no urgency to remove supers, extract, clean up, and store supers, but it is best to begin between mid-August and early October. After the extracted supers have been cleaned up and stored, begin the medicating and feeding (if you feel that feeding is necessary). If you are going to feed, medicate for Nosema Disease. Remember that in the higher elevations, even in the far South, winters may be severe enough that you should consider yourself to be in the North.

If you are going to re-queen in the fall, remove your honey supers about the last week in August, and have your new mated queen arrive about that time. Most queen producers stop producing queens about early September. You can re-queen later if you can find a producer in the South who banks queens into the winter. Be sure to de-queen before re-queening. (See Unit 13 for steps on re-queening.)

If you live farther north in the southern area, you will have less time to get the medication and feeding done between the end of the honey flow and the start of colder weather. As in the North, feeding tends to stimulate brood rearing, which drops off when the honey flow ends. Young bees are an important winter requirement in the South as well as in the North.

A word of caution for those who live in areas where there is a winter flow from pepper trees, eucalyptus, etc.: In those areas, plan to have the medication completed before the winter flow begins and supers are added for the flow. Nosema and AFB medication will keep brood rearing strong and the larvae healthy, making more bees available for the winter flow. Be sure to inspect the honey supers and add them as needed to avoid crowding and early swarming. Do not medicate while the winter flow is on.

In areas with no winter flow, half of the Nosema medication can be fed in the fall to improve the wintering condition, and the other half can be fed over an extended period through the winter (say, a quart every two weeks through the winter until the suggested three gallons are used up). Nosema medication is effective if it is in the food of the adult bees during the winter when the Nosema protozoa are growing in the midgut of the bee. In the North, two gallons must be fed in the fall so the medication is available in the food stored and used by the cluster. In the South, where the bees are active on most days, feed the Nosema medication throughout the winter to be sure it is available when the disease is more active in the less-active bees.

FALL MEDICATING TO CONTROL AFB/EFB

If you mixed a year's supply of the Terramycin®-powdered sugar/Drivert® sugar mix in the spring, you should have enough left for the two medications you will make in the fall. If you must mix the AFB medication at this time, refer to Unit 11, "Preparing the AFB/EFB Medication." Inspect the brood chamber to check on your queen. Refer to Unit 12 for the steps to inspect the brood chamber.

Equipment needed:
- [ ] Hive tool
- [ ] Smoker or hive bomb
- [ ] Frame grip
- [ ] Empty super
- [ ] AFB/EFB-medicated mix
- [ ] Tablespoon

Step-by-step directions:
- ■ **Dress properly.**
- ■ **Light the smoker** or have hive bomb handy.
- ■ **Enter the hive.** Place the inverted hive top on the ground nearby. Set the empty super on the inverted hive top.
- ■ **Inspect the brood chamber** (see Unit 12). When inspection is complete, go to the next step.
- [ ] Place seven tablespoons of the AFB-medicated mix on the center frame tops of the upper brood chamber by drawing the spoon along the top of the frame. Try to keep the mix on the back two-thirds of

the frame tops. It will not hurt if some of the mix drops into the hive. Close the hive.
- [ ] Repeat the previous medication steps above in ten to fifteen days, omitting the inspection of the brood chamber if you wish.

FALL FEEDING TO CONTROL NOSEMA DISEASE

In the North: This feeding of Nosema-medicated sugar solution will probably take place at the same time that you are medicating to control AFB. The sugar solution used in this medication is the 2-1 sugar solution; to prepare this solution refer to Unit 11, "Preparing the Sugar Solution for Medicating and Feeding." To prepare and dispense the Nosema-medicated sugar solution, see Unit 11, "Medicating to Control Nosema Disease."

In the South: A separate set of steps are given in this section, directing you to feed the Nosema-medicated solution throughout the winter.

Equipment needed:
- [ ] Nosema-medicated sugar solution
- [ ] Entrance feeder

In the North

Step-by-step directions:
- [ ] Place the feeder base in the hive entrance.
- [ ] Fill a quart jar with the Nosema-medicated solution. Screw on the cap securely, and invert the jar over and into the feeder base.
- [ ] As the jar empties to about one inch from the feeder cap, remove, refill, and replace the jar. **Caution:** Use gloves when unscrewing the feeder cap so you will not be cut by the sharp edges.
- [ ] Repeat the previous step until two gallons of the Nosema-medicated sugar solution have been fed.

An alternate method that you may find to be as good or better is this: Rather than feeding externally with the entrance feeder, feed internally by setting the entrance feeder over or next to the oval hole in the inner cover with an empty super over the feeder with the cover on top. If you want to feed faster, punch holes in the lids of peanut butter jars or in the plastic lids of coffee cans, and fill those containers with solution. Place each container on a pair of twigs on the frame tops of the brood chamber. The twigs allow the bees to feed from underneath. In this case, you can put two or more containers of feed on the hive directly over the bees (inside an empty super) for a more rapid feeding if colder weather is expected soon. Internal feeding is best if there may be only a few hours during the day when the temperature is above 55° F. This method can also be used in the spring if you want to feed faster for any reason.

In the South

Step-by-step directions:
- [ ] Place the feeder base in the hive entrance.
- [ ] Fill a quart jar with the Nosema-medicated solution. Screw the cap on securely, and invert the jar over and into the feeder base.
- [ ] Feed one quart of the medicated sugar solution each week for four weeks.
- [ ] Feed one quart every two weeks until two more gallons of medicated sugar solution have been fed. Feed a total of only three gallons.

If you begin feeding in October, you will keep the medicated solution in the food stores until early February. This is the time when the Nosema medication can most effectively control Nosema Disease. If this schedule appears to be too time consuming, use the schedule for "In the North." If your area has a winter flow, medicate as in the North, but stop medicating several days before adding honey supers for the winter flow.

FALL FEEDING OF A NONMEDICATED SUGAR SOLUTION

Use the 2-1 sugar solution to get the hive into a better wintering condition. (See Unit 11 for the steps to prepare the 2-1 sugar solution.) The steps for feeding the nonmedicated sugar solution are the same as for feeding the medicated sugar solution. In both areas, North and South, follow the steps in the previous section for "In the North."

Feed sugar solution until you think the hive is well stocked or until the bees are no longer active. The sugar solution can be fed internally, as was mentioned earlier, if colder weather seems to be coming on faster than expected or if you want to get food stores to the hive quickly for any reason.

UNIT **24**

Uniting Weak Hives

If two hives are placed together or are united directly, the bees will fight among themselves, and many thousands will die. There is a method of uniting hives that permits the beekeeper to unite two hives without causing fighting, and it is almost 100 percent sure. Very simply, the method is to put a sheet of newspaper between the brood chambers of the two hives before putting them together or uniting them. Uniting hives with the newspaper method solves a number of beekeeping problems, such as laying worker hives, queenless hives, or weak hives. (See Figure 24-1.)

Two strong hives can be united to cut down the number of hives in your apiary; however this is usually not the reason for uniting hives. In general, it is a fall manipulation to unite a weak hive that will not make it through the winter with a strong hive. Store the extra brood chambers until spring, when they can be restocked with a package or swarm. In this case, uniting allows you to utilize the live bees from the weak hive and eliminate the clean-up of a dead hive in the spring.

A weak hive is one that has not developed in population as it should or may have gotten started too late and does not have sufficient stores to get through the winter—or both. If the thought even crosses your mind that a hive will not make it through the winter, that is reason enough to unite it with a strong hive.

Uniting may be done at any time of the year. Some other times that uniting could or should be done are the following.

1. If you find a laying worker hive and swarms are not readily available, unite the laying worker hive as soon as possible with a strong hive in your apiary.
2. If you find a hive with the very first signs of wax moth infestation and if swarms are not readily available, unite the wax moth infested hive with a strong hive in your apiary. In most cases, wax moth infested hives also have laying workers.

Figure 24-1, Uniting a Weak Hive with a Strong Hive. A single sheet of newspaper is placed over the brood chamber of the strong hive. After taping the paper down (see black areas), punch about six holes in it with a hive tool, pencil, knife, or nail. Set the weak hive brood chamber on the newspaper, and the hive is united.

3. If you find a queenless hive in the later part of the honey flow or just after the honey flow, it should be united with a strong hive unless the population is very strong, a mated queen is immediately available, and the hive has a considerable store of honey and pollen. Most beekeepers hope against hope that the weak hive will make it through the winter, and it usually will not. Experience and the loss of weak hives will prove the truth of this statement.

Weak hives should be reduced to a single brood chamber before uniting. A brood chamber that is united with no queen or that has been de-queened can be united above the queen excluder, and the uniting procedure is simplified.

It is best to unite the weak hive with the nearest strong hive. Field bees from the weak hive will return to the old location after the newspaper has been torn away. If the united hives are setting side by side or only a few feet apart, the field bees from the united hive will find their way into the new hive by dark.

Remember: You will not gain anything by uniting a weak hive with another weak hive. Two weak hives do not make a strong hive. Unite a weak hive with a strong hive. There are times when uniting can be made directly, without the newspaper, but one does not know when those times are without lots of experience. It is not that much trouble to slip a piece of newspaper between the hives being united to make sure that no fighting will occur. Only in cases where the population is very small and laying workers or the first signs of wax moth are present should uniting take place without the newspaper.

If your weak hive is queenless, there will be less manipulation in the uniting procedure. Read through the two uniting methods, and decide whether to de-queen before uniting. If you have laying workers or the first signs of wax moth, you can be positive that the hive has no queen, so use the method for uniting a queenless hive. If your hive is queenless or you have de-queened your hive, use the method for uniting a queenless hive. If you find eggs and larvae in the hive to be united and you do not de-queen the hive, or if you find eggs and larvae and cannot find the queen, then use the method for uniting a queenright hive with a strong queenright hive.

If two queenright hives are united, the bees or the meeting of the two queens will decide which one will become the mother of the hive. In general, the youngest queen will become the mother of the hive if the choice is made between the two queens and not by the bees themselves. If the bees make the choice, the hive with the strongest population will usually destroy the queen from the weaker hive.

In the South: You have a better chance of bringing a weak hive through the winter than beekeepers in colder climates, because you can feed and inspect the hive through the winter. However, unless it is your only hive, it will be less trouble and less expensive to unite the weak hives, store the brood chambers with PDB moth crystals, and restock the hive in the spring with a swarm (or several swarms). Also, you do not run the risk of losing the comb to wax moths.

UNITING A QUEENLESS HIVE WITH A STRONG QUEENRIGHT HIVE

Before beginning, get the bees in the weak hive into one brood chamber by shaking them from the second brood chamber. This should be done the day before uniting. Store the second brood chamber in a plastic bag with PDB moth crystals. If you are going to remove the queen from the weak hive, do it *just an hour or two before* uniting the two hives.

Equipment needed:
☐ Hive tool
☐ Smoker or hive bomb
☐ Frame grip

UNITING WEAK HIVES   247

- [ ] Sheets of newspaper
- [ ] Roll of masking tape

After three days or three weeks:
- [ ] Extra top or pair of two-by-fours
- [ ] Inner cover
- [ ] Bee escape

Step-by-step directions:
- ■ **Dress properly.**
- ■ **Light the smoker** or have hive bomb handy.
- ■ **Enter the strong hive.** Smoke, then set any honey supers on the inverted hive top.
- ■ **Inspect the brood chamber** if you have not already done so. (See Unit 12.)
- [ ] Replace the queen excluder if you inspected the brood chamber. Leave it in place if you did not inspect the brood chamber. Bring out and place a queen excluder over the upper brood chamber of the strong hive if no queen excluder was on the hive.
- [ ] Place a sheet of newspaper over the queen excluder. Tape the newspaper to the sides of the brood chamber so the paper will not blow off. Punch several holes in the paper with a hive tool, pen knife, pencil, nail, etc.
- [ ] Move to the weak hive that is to be united on the strong hive. If you know that the hive is queenless, do not remove the top. Lift the brood chamber from the bottom board, and set it on the newspaper that is on the strong hive.
- [ ] If you removed honey supers from the strong hive, remove the hive top of the united hive, and place those supers on the united hive. Replace the hive top.
- [ ] *Wait three days* if the united brood chamber was queenless and contained no brood. If the united brood chamber was de-queened and contained brood, *wait three weeks until the brood emerges,* and then proceed with the following steps.
- ■ **Dress properly.**
- ■ **Light the smoker** or have hive bomb handy.
- [ ] Smoke, then remove any honey supers on the hive. Set them on the inverted hive top or on the pair of two-by-fours.
- [ ] Smoke between the united brood chamber and the queen excluder. Wait one minute, and then set the united brood chamber on the inverted hive top or pair of two-by-fours. Remove any of the newspaper remaining on the queen excluder.
- [ ] Replace any honey supers on the queen excluder. Place the inner cover with the bee escape on top of the topmost honey super. If there

were no honey supers, place the inner cover with the bee escape on the queen excluder.
☐ Set the united brood chamber you just removed back on the inner cover. Close the hive.
☐ *Wait two days.*
■ **Dress properly.**
☐ Remove the united brood chamber from the hive, and store it in a plastic bag with PDB moth crystals.
In the North: If winter is coming, you can store the brood chamber in an unheated, bee-free area, protected so that mice cannot get into the brood super during the period of storage. If you are not going to put the brood supers back in service in the spring, store them in a plastic bag with PDB moth crystals.
☐ Remove the bee escape. Close the hive.

UNITING A WEAK QUEENRIGHT HIVE WITH A STRONG QUEENRIGHT HIVE

The following steps will also apply to the uniting of a strong queenright hive with another strong queenright hive, should that ever become necessary, with the exception that both brood chambers will have to be united. These steps discuss uniting a weak queenright hive with a strong queenright hive.

Within several days of uniting, bring the weak hive to be united down into a single brood chamber. Do this by shaking the bees from the brood chamber with the least amount of brood into the remaining brood chamber. Exchange any frames of brood in the brood chamber being removed for empty or nearly empty frames. Store the brood chamber that was shaken free of bees in a plastic bag with PDB moth crystals.

Equipment needed:
☐ Hive tool
☐ Smoker or hive bomb
☐ Frame grip
☐ Sheets of newspaper
☐ Roll of masking tape
☐ Bee brush

After three weeks:
☐ Inner cover
☐ Bee escape
☐ Pair of two-by-fours or extra top
☐ Empty super

UNITING WEAK HIVES 249

Step-by-step directions:
- ■ **Dress properly.**
- ■ **Light the smoker** or have hive bomb handy.
- ■ **Enter the strong hive.** Smoke and remove the honey supers, if any. Place them on the inverted hive top.
- ☐ Puff smoke over the queen excluder and gently remove it. Stand it nearby.
- ☐ Place a sheet of newspaper over the upper brood chamber of the strong hive. Tape the sides of the newspaper to the brood super so the paper will not blow away. Punch several holes in the newspaper with a hive tool, pen knife, pencil, etc.
- ☐ Move to the weak hive. Do not remove the hive top. Lift the brood chamber from the bottom board, and set it on the newspaper that is on the strong hive. Check that the queen excluder is not on the hive but that it has been removed as directed above.
- ☐ Remove the top from the weak hive if you removed honey supers; otherwise, go to the next step. Place the queen excluder on top of the weak hive's brood chamber that has just been placed on the strong hive. Replace the honey supers, if any. Close the hive.
- ☐ *Wait three to four days.*
- ■ **Dress properly.**
- ■ **Light the smoker** or have hive bomb handy.
- ■ **Enter the hive.** Smoke and remove the honey supers, if any. Place the honey supers on the inverted hive top or on the pair of two-by-fours.
- ☐ Smoke between the united brood chamber and the upper brood chamber of the strong hive. Wait one minute. Set the united brood chamber onto the extra hive top or on the pair of two-by-fours. Smoke over the queen excluder and gently remove it. Stand it in a safe place nearby.
- ☐ Remove the remnants of newspaper from the hive. You are now going to shake the bees from the united brood chamber to make sure the queen or queens are in the brood chambers of the strong hive.
- ☐ Shake the bees from each frame of the united brood chamber back onto the frame tops of the upper brood chamber. (See Unit 13.) Stand the shaken frames nearby, or place them in an empty super.
- ☐ When all frames in the united brood chamber have been shaken and the inside of the brood chamber has been brushed free of bees, return all the frames to the brood chamber. Place the queen excluder on the upper brood chamber of the strong hive, and set the united brood chamber, which may contain brood in all stages, on

the queen excluder. Replace the honey supers, if any, and close the hive.
- ☐ *Wait three weeks* for the brood to emerge from the frames in the united brood chamber.
- ■ **Dress properly.**
- ■ **Light the smoker** or have hive bomb handy.
- ■ **Enter the hive.** Smoke and remove the honey supers, if any. Set them on the inverted hive top or on the pair of two-by-fours.
- ☐ Smoke between the queen excluder and the united brood chamber. Wait one minute. Set the united brood chamber on the inverted hive top or on the pair of two-by-fours.
- ☐ Replace the honey supers, if any. Place the inner cover with the bee escape in place on top of the topmost honey super, if any. If there are no honey supers, place the inner cover with the bee escape in place on the queen excluder.
- ☐ Set the united brood chamber on the inner cover, and close the hive.
- ☐ *Wait two days.*
- ■ **Dress properly.**
- ☐ Remove the united brood chamber, and store it with PDB moth crystals in a garbage bag. *In the North:* You can store the removed brood chamber in an unheated, bee-free area until spring, if you are going to restock the hive (and it is now fall).
- ☐ Remove the bee escape. Close the hive.

UNIT 25

Wintering the Hives

Beginning beekeepers will be well advised to talk to experienced beekeepers in their areas. Follow the procedures for wintering practiced by those who have had good luck in bringing their hives through the winter.

Preparation for winter begins with the bees' first spring activity. Everything you do during the season should lead you to the end of the season with a good population of young, healthy bees and an adequate store of pollen and honey to carry the hive into the next spring. An adequate amount of honey and pollen stores should be left in the spring to feed the ever-increasing brood build-up until the bloom of the earliest spring flowers and fruits can provide adequate nectar and pollen to continue the population growth. Early spring is one of the most critical periods in the life of the hive, because the hive may run out of stores and starve just before the early spring flowers can save it. This is the reason to heft the hive more often in the early spring. If the hive lightens considerably, it will be necessary to feed the hive until nectar and pollen are available in the field.

The prime requirements for successfully wintering a hive in either the North or the South are the following.

1. A large population of young, healthy bees.
2. Adequate stores of honey and pollen.
3. Venting of the hive to expell water vapor. Water vapor in any form condenses and drops among the bees. It is generated by the metabolic processes of the bees and by water condensing from the humidity of the air near the cluster.
4. An auxiliary exit kept free of snow, ice, frost, and debris. The auxiliary exit is important in case the hive entrance becomes clogged by ice or snow (in the North) or by a mass of wet leaves (in the South).

In most cases, the venting arrangement provides both venting and an auxiliary exit.

Warmth is not a requisite for the wintering of bees. The bees cluster in cold weather; the colder the weather becomes, the more tightly the cluster draws together to form an insulating layer of bees on the outside of the cluster. Inside the cluster, under the insulating layer, the bees are not packed so tightly together and are capable of feeding on the honey stores within the cluster and shaking or quivering to produce heat. Feeding in the cluster is done by the bees in the center passing food along to those nearby, who, in turn, pass the food along to those bees farther from the center. Unless the weather is unusually cold for a long period of time, the bees in the insulating layer gradually work their way back inside the cluster. In very severe weather and over the period of winter, many bees die of exposure on the outer edges of the cluster, and the old bees die naturally. This is one of the reasons for the auxiliary exit, for sometimes there may be so many dead bees that the hive entrance becomes plugged.

During the winter it is important for there to be an occasional pleasant, calm day on which the sun shines and the wind does not blow. On those days, the hive will warm sufficiently inside that the cluster can break somewhat and the bees can move to a new area of stores. The bees will then be ready for another month of cold, windy weather. During winters when the weather is blustery and cold for many weeks without a break, the hive may die from starvation, because the cluster cannot break and move to new stores. This is known as *winter kill* and the beekeeper can do nothing to prevent it. In winter kill you will find the hive dead with stores of honey still available around the cluster that has starved to death. An individual bee will die of cold at about 43° F, yet the bees in the cluster can stand very cold weather for short periods of time.

WINTERING HIVES IN THE NORTH

It is a good idea for the hobbyist to wrap or box the hive for the winter. Wrapping is done with the black felt or tar paper that is used in the building trade. Boxing is done with a black waxed box, available from bee supply houses, which will fit down over a two-brood-chamber hive. For the hobbyist with a small number of hives, the waxed box is the best investment of time and money. The black covering on the hive helps absorb warmth on those few pleasant, calm days during the winter. It is not recommended that the hive be packed or insulated in any way. An insulated hive will keep the warmth of the sun from getting through to warm up the hive on the pleasant days and may cause winter kill from intended kindness. Again, do not try to keep the bees warm, for they will do that in the cluster (if stores are available).

When the hive is wrapped or boxed, venting is usually accomplished by boring a ¾-inch hole in one end of the upper brood super under the handhold, or near the side of the brood chamber about two inches from the top of the super. The wrapping or box is cut away, and the material is tacked or stapled securely around the hole so it will not tear during winter storms. In wrapping the hive, be particularly careful that the wrapping over the top of the hive is tied down securely so water will not leak into the hive during rains or melting snows. When you use the waxed box, the hive top may not fit down over the box; it must be fastened securely so that it will not blow off during winter storms.

Some authorities suggest that brood chambers be arranged so that the brood chamber with the brood and most of the bees is placed on the bottom board and the other brood chamber of stores is placed as the upper brood chamber. This manipulation has some merit because it allows the bees to move upward to new stores as winter proceeds.

In some of the northern areas with the most severe winters, a third super of food stores should be left on the hive for winter stores. If winters are so severe that you lose your hives as often as they are brought through, consider restocking the hive every spring with a package. Extract most of the honey from the brood chambers, and plan on restocking a dead hive every spring.

If you have a number of hives, you might try to unite two strong hives in the fall and carry those super-strong hives through the winter with three of the four brood chambers packed with stores. In the spring, restart each of the united hives by borrowing brood and introducing a new mated queen, then bolstering those hives with a swarm, when available. Remember, one live hive in the spring—even a weak one—is better than two dead hives. See Unit 24 for a discussion on uniting hives.

If you have more than two brood chambers on the hive, you cannot use the waxed box for a winter cover. In this case, use a round or two of tar paper fastened securely at about six-inch intervals, from top to bottom, with stout twine. The top must be folded over and fastened most securely so that moisture cannot get into the hive. You might want to tie a cap of canvas or plastic over the top of the hive before tying down the hive top. Remember to keep the auxiliary exit/vent open and the wrapping around it tacked down securely.

Before wrapping the hive, you might want to try some old tricks for absorbing interior-generated moisture: a rectangle of carpeting cut to fit over the frame tops of the topmost super, or a super over a screen on the topmost super filled with straw or other absorbent material. The carpeting has an advantage over an inner cover in that moisture is absorbed and does not drop down on the cluster. Do not use these tricks in place of the vent recommended, but use them in conjuction with the vent.

In most areas today, hives are not wrapped for the winter. Deciding whether to wrap must be up to you, because you want the best for your bees over the winter. In some cases, hives can be moved together and wrapped in pairs or quadruples.

Odds and Ends

Do not forget to place the entrance reducer in the hive for the winter months. This may cut down some drafts, but it is more important to keep mice out of the hives while the bees are clustered during the winter. Mice can cause quite a lot of damage to the comb.

Heft the hive occasionally, particularly in late winter. If the hive begins to lighten considerably, be prepared to feed the hive as soon as the bees can become active. Unite weak hives before winter sets in. (See Unit 24.)

Do not leave empty honey supers on the hive during the winter, but store them in an unheated, bee-free area. Store the supers so mice cannot get into them from above or below.

Emergency Feeding in Late Winter or Very Early Spring

Make your final brood chamber inspection in the fall, and feed the bees if necessary. Always remember that honey is the best food for bees, and you can add frames of honey in the comb to those hives that need help going into winter. You may want to hold back a super or two of honey in storage or in the freezer to have it on hand for early spring feeding. If you have only one or two hives, hold back two or three frames of comb honey for spring feeding.

A quick way to feed bees that are in a bad way is to make up a sugar solution and use a mist sprayer to spray it into the frames of empty combs. Make up several combs with the sugar solution sprayed into both sides of the comb, and place them in the hive up next to the cluster. *Do not break into the cluster!*

It is possible to feed the bees granulated sugar on the inner cover, rim side up. Pour the sugar around the opening of the inner cover; if they are not too weak, the bees will be able to get by until warmer weather, when you can feed sugar solution on top of the inner cover hole or through the entrance. If you feed combs of honey to the bees and the bees are not very strong, tear the cappings to open the cells so the bees will not have to do extra work to get to the honey.

WINTERING HIVES IN THE SOUTH

Prime wintering requirements are as important in the South as in the North. In most southern areas, the hives need not be wrapped or boxed. In the more northern parts of the South you probably will not have more than four to six weeks of intermittent cold, rainy, windy days through the winter, with perhaps a chance of snow lasting not more than a week.

Venting of the hive in southern areas is important, because in most areas, even though the temperature during the day may be in the 60's or 70's, the nightly temperatures may drop into the 40's or 50's, creating condensation of water vapor in the hive. Rather than boring holes in the upper brood chamber, as is necessary in the North, insert a ¼-inch square block of wood that is one inch long under the front edge of the inner cover on each side of the brood chamber. A one-inch length of pencil will do very nicely. Because the bees can get through any opening greater than ³⁄₁₆ inch, this vent will also act as an auxiliary exit. If you worry about the opening along the length of the inner cover, you can close those openings with masking tape. For the auxiliary exit to work, place the hive top on the hive and push it all the way forward, leaving room for the bees to come out the vent and down under the forward lip of the hive top, the same way that the water vapor escapes from the hive. If you have a flat top on your hive and no inner cover, place the blocks under the end of the flat hive top and push it all the way forward, or make your vent come out the side of the hive top and tape off all but about four inches. (See Figure 25-1.)

Figure 25-1, Winter Venting in the South. The inner hive cover is being held up from the brood chamber on the front end by a short length of a ¼-inch square of wood on each side. Short pieces of pencil or ¼-inch doweling would work nicely. The cover should be placed on the hive and pushed forward as far as possible to make an auxiliary exit for the bees.

Close the hive entrance around Thanksgiving with an entrance reducer. This cuts down on drafts and, more importantly, keeps mice out of the hives that become weak or die during the winter. Remember that mice can slip through a ¾-inch opening very easily, so the entrance should be blocked to a 3- by ⅜-inch opening, which is most easily done by buying or making an entrance reducer with the proper length that will fit into the ¾-inch entrance of your hive.

Heft the hive during the winter, particularly in late winter and very early spring, to see if the hive needs additional feeding. The emergency feedings suggested for hives in the North will apply to hives in the South as well. The early spring build-up is the most critical time in the wintering of hives in the southern areas. If you have been medicating through the winter to control Nosema Disease, or if you medicate to control Nosema Disease early in the spring, you will be adding food stores in this critical period.

If you feel the hive is not doing well, inspect on any day when the temperature is above 60° F. Pick a pleasant, sunny, calm day, and inspect the brood chambers, check the honey stores, and look for signs of AFB.

You can unite weak hives during the winter in the South. (See Unit 24.) If you find a laying worker hive or a hive with the first signs of wax moth infestation, unite it with a strong hive, even during the winter. In most winter uniting, the hives to be united are very weak, and it is recommended that the hives be united without newspaper. In most cases, the combs at that time are worth more than the bees in the weak hive.

Do not leave the empty honey supers on the hive over winter. If you have a winter flow in your area, medicate the hive, and then get a honey super out of storage, air it for a day or two, and put it on the hive over a queen excluder. If you do not have a winter flow in your area, store all the honey supers with PDB moth crystals until needed next season.

PART **VII**

Miscellaneous Activities and Manipulations

The next ten units cover activities or manipulations that may be required on various occasions as you continue to keep bees.

- **Unit 26** LAYING WORKERS
- **Unit 27** WAX MOTH INFESTATION
- **Unit 28** REMOVING BEES FROM HOUSES, TREES, ETC.
- **Unit 29** REMOVING BEES WHEN THE COMB CAN BE EXPOSED
- **Unit 30** REARING A FEW QUEENS FOR YOUR OWN USE
- **Unit 31** RE-QUEENING A HIVE OF MEAN BEES

258　KEEPING BEES

- **Unit 32**　ROBBING
- **Unit 33**　THE OBSERVATION HIVE
- **Unit 34**　POLLEN TRAPS AND THEIR USE
- **Unit 35**　THE SOLAR WAX MELTER
- **Unit 36**　LOOKING BACK AND LOOKING AHEAD

UNIT 26

Laying Workers

There is always a chance that the virgin queen generated within a hive from a young larva may be lost on her mating flight. If this happens, the hive will become hopelessly queenless, because there are no young larvae from which to generate a new queen. The only help for this hive would be the intervention of a beekeeper, who could furnish new brood or a new mated queen.

If the beekeeper does not discover that a hive is hopelessly queenless, within about four weeks after the queen disappears, one or many worker bees will begin to lay eggs in the cells. Laying workers do not mate and do not carry sperm, so all the eggs they lay are infertile and will become drones or male bees. Perhaps the lack of queen substance after the queen disappears causes some worker bees who have never tasted queen substance to begin to lay. Queen substance inhibits egg laying in workers when the queen is present. Situations that can lead to laying workers in the hive are:

- Hives that have swarmed and are not checked after at least four weeks or by the start of the honey flow, which is about the time the second honey super is added.
- A hive or hives that are worked carelessly or roughly where the queen may be killed or dropped from the hive.

In both cases, new queens will be generated; if the queen is lost on the mating flight, the hive is done for if the loss is not discovered.

Fortunately, signs of laying workers are very easy to spot. Because the abdomen of the worker bee is not long enough to reach the bottom of the cells, the eggs will fall out of the laying worker. Until she feels a pressure as the egg is laid, she will continue to drop eggs into a cell. If you are inspecting the brood frames and spot cells with multiple eggs in them, you can be sure that the hive has laying workers. Another clue is that the laying workers do not lay in a neat pattern as the queen does. Laying workers lay eggs in a hit-and-miss pattern that is quite noticeable when the brood is covered, because you can see the extended drone cappings scattered around on the frame. Drones being reared in worker cells are another indication of laying workers. Drone-laying queens will also lay in worker cells, but here again, the queen will lay in a neat pattern, and the drone cappings will not be scattered as they are with laying workers. (See Figure 26-1.)

Laying worker hives are very difficult to re-queen by the normal procedure. Apparently, bees in the laying worker hive think they have a queen even though they continue to try to generate new queen cells from the drone larvae. As a result, they tend to destroy any new mated queen that is offered to the hive.

In past years it was thought that by moving the hive some distance away and shaking all the bees, the laying workers would not return to the hive and a new queen could be given to the hive. Because this procedure of moving and shaking the bees seldom worked, it is not surprising that recent research has shown that laying workers can fly and will return to the hive to continue to lay. Those of us who thought that moving and shaking might somehow work can now forget it as a possible way to re-queen a laying worker hive.

What you must do is introduce a quantity of bees and a queen to the laying worker hive. Some ways to add bees to the hive are:

- Add a swarm or two, if swarms are available.
- Add several frames of brood and bees from a strong queenright hive, and add a new mated queen.
- Add the frames from a nucleus hive with its own queen into the laying worker hive.
- Unite the laying worker hive with any moderate-to-strong hive in the apiary.
- As a last resort, kill the bees by placing each of the brood supers, bees and all, in plastic garbage bags with PDB moth crystals, tying the bags securely, and starting the hive again next season.

Remember, the time of year you discover the laying worker hive will determine which method to use in restoring the hive to productivity or taking it out of service, particularly in the North. If laying workers are discovered during the swarm season, it is best to restore the laying worker hive with a swarm or several swarms. If laying workers are discovered after about half the honey flow is over, it is best to unite the laying worker hive with one of your strong hives. If you have only one hive, it is best to place the supers, bees and all, in plastic bags with PDB.

LAYING WORKERS 261

Figure 26-1, Capped Drone Brood. Top: A good example of the hit-and-miss laying pattern of laying workers. This pattern is easy to spot when the drone larvae have been capped in the laying worker hive, but it can also be seen when the eggs and larvae are in the cells before any cells have been capped. Bottom: Capped drone cells as they occur around the edges of the frames in a queenright hive. These drone cells are along the bottom bar. In both photos, notice how the drone cappings extend beyond the brood comb with a bullet-shaped capping.

In the North: If laying workers are discovered around the first of April (before swarm season), and you have several strong hives, you can order a new queen, borrow about four frames of brood and bees, and re-queen the laying worker hive when your new mated queen arrives. This may work just after swarm season is over if you have some strong hives. Borrow two frames of bees and brood from each of two hives, and re-queen the laying worker hive.

In both the North and South, you must consider whether the restored hive can get itself into strong enough condition to make it through the winter. It does not pay to go to the trouble and expense to restore any hive if it must be united in the fall. You must make the decision.

WHAT TO DO WHEN YOU HAVE LAYING WORKERS IN YOUR HIVE

If You Have Only One Hive

If you can capture a swarm within two weeks, add it to the hive. (See Unit 16 for capturing swarms and Unit 17 for adding a swarm to re-queen a hive.) If it will be longer than two weeks, place each of the brood chambers, bees and all, in a plastic bag with PDB crystals until you can restart the hive in the spring.

If You Have Two or More Hives

You can perform the procedures for one hive or use any of the methods that follow.

- Unite the laying worker hive with one of your moderate to strong queenright hives. (See Unit 24.)
- Add three or four frames of brood and a new mated queen into the laying worker hive. Before starting this procedure, you must order and have the new mated queen in hand. You will be borrowing frames of brood from one or several hives.

Equipment needed:
- ☐ Hive tool
- ☐ Smoker or hive bomb
- ☐ Frame grip
- ☐ Mated queen in cage

Step-by-step directions:
- ■ **Dress properly.**
- ■ **Light the smoker** or have hive bomb handy.

LAYING WORKERS 263

- ■ **Enter the laying worker hive.**
- ■ **Inspect the laying worker hive** and remove any frames with brood—eggs, larvae, or capped drone cells. Shake the bees from these frames, and stand the frames in a safe place.
- ☐ Place the brood frames removed from the laying worker hive in a bee-free area for two days so that the brood will chill and die. As soon as possible, get these frames back into service in a honey super above the queen excluder so that the bees can clean out the dead brood.
- ☐ Enter the hive or hives from which you are going to borrow brood and attached bees. Borrow three to five frames of brood from one or several of your strong hives. Check the borrowed frames carefully to make sure you do not transfer the queen with the brood.
- ☐ Place the frames of borrowed brood into the center of the laying worker hive. Replace any borrowed frames with frames of comb or foundation. Be sure you have replaced the proper number of frames for the frames of borrowed brood.
- ☐ Close the hive or hives from which the brood was borrowed.
- ☐ Introduce the new mated queen on the borrowed frames of brood and bees that are in the laying worker hive. (See Unit 13 for steps on re-queening the hive.) In this re-queening, you should perforate the candy plug.
- ☐ Check the hive after four days to make sure the queen has been released; if not, release the queen manually.
- ☐ Check the hive one week later to make sure the queen is laying. Add a second brood chamber and honey supers as required during the season.
- • Add a nucleus hive or a recently started swarm into the laying worker hive.

Equipment needed:
- ☐ Hive tool
- ☐ Smoker or hive bomb
- ☐ Frame grip
- ☐ Nucleus hive

Step-by-step directions:
- ■ **Dress properly.**
- ■ **Light the smoker** or have hive bomb handy.
- ☐ Move the nucleus hive alongside the laying worker hive, or if the nucleus hive is stronger than the laying worker hive, move the laying worker hive alongside the nucleus hive. **Enter both hives.**
- ☐ While inspecting the brood chambers from the laying worker hive, remove any frames containing brood—eggs, larvae, and capped

drone cells. Shake the bees from these frames, and stand the frames in a safe place.

- ☐ Remove all the frames from the nucleus hive, one by one, and place them in the center of the laying worker brood chamber. Dump and brush any bees on the inside of the nucleus hive onto the frame tops of the frames just added to the laying worker hive.
- ☐ Close the hive. The field bees from the transferred hive will return to the old location. This is why you move the weaker hive to the stronger hive.
- ☐ Place the brood frames removed from the laying worker hive in a cool, bee-free area for two days until the brood has chilled and died. As soon as possible, put these frames back in service in a honey super over the queen excluder so the bees can clean out the dead brood.
- ☐ Inspect the hive after four days to make sure the queen is continuing to lay. Add a second brood chamber and honey supers as required during the season.

Quite honestly, a laying worker hive is not worth wasting much time on. If it can be salvaged during swarm season, it is worth the time; otherwise bag it with PDB or unite it with another hive.

If laying worker hives are not restored to productivity or done away with by placing the brood chambers, bees and all, into plastic bags with PDB moth crystals, they will soon be attacked and destroyed by wax moths. Avoid the expense and mess of losing your comb this way. Wax moth infestation is discussed in the next unit.

UNIT 27

Wax Moth Infestation

The wax moth is probably in every hive in the world. The adult wax moth slips into the hive and lays its eggs. If the hive is strong, the bees capture and carry out the wax moth larvae as they are hatched, dropping them some distance from the hive. A lot of the larvae must live after being carried from the hive, because there are always more adult moths to enter the hives and lay more eggs.

Foundation wax can be stored for many years without being attacked by the wax moth. The wax moth is instinctively smart enough not to waste its eggs in an area where the larvae have no food stores to grow and reproduce, which is what the game is all about. White honey comb will not be destroyed as fast or as completely as dark brood comb. The wax moth larvae must have protein to develop, and this is furnished by the pollen, propolis, and cocoon materials in the brood cells and the propolis around the edges of the honey frames.

You are probably aware that wax moth infestation follows laying workers, which result when the hive is hopelessly queenless. Because no worker bees are produced as older worker bees die, the queenless hive has fewer and fewer bees to clean out the wax moth larvae that begin to develop faster and faster. As long as a hive is queenright and strong, you do not have to worry about wax moths, except in brood chambers or honey supers not on those strong hives or not stored under the protection of PDB moth crystals.

Any super removed from the hive and placed in a warm area for a period of several weeks will have wax moth larvae and adult moths. If the super is open to the light, wax moths do not lay eggs so soon as they do if the super were closed and dark. If the weather is cold or cool, wax moth larvae do not develop so fast as they do if the weather is quite warm. Wax moth eggs do not develop into larvae when the temperature is below about 45° F. The moth eggs and tiny larvae in comb honey can be killed if you freeze the comb for several days. Honey supers stored in unheated, bee-free sheds in the North do not have to be protected by wax moth crystals, because they are protected by the cold as long as the temperature is below 45° F.

FIRST SIGNS OF WAX MOTH INFESTATION

If you are lucky enough to catch wax moth infestation in the early stages, you will first see fine webbing over several cell tops, perhaps in a number of areas of the comb. At this stage, the wax moth larvae are very small and will grow very rapidly, depending on the temperature and the food available.

If you are a little less lucky, you will next find that the larvae are larger and are tunneling through the comb along the midrib. As the larva moves along in the comb, it spins a web tunnel so that it cannot be so easily captured by and carried from the hive by the worker bees. Once the wax moth has proceeded so far as to make tunnels along the midrib of the comb, you can be absolutely sure that you have wax moth infestation; if something is not done *immediately* your comb will be destroyed in a matter of days.

If your luck is really bad—and for most of us it is—you will not discover the wax moth infestation until both the brood supers and possibly a honey super or two are totally involved in the infestation. At this stage, there are masses of webbing and cocoons all through the brood combs. As they spin their cocoons before pupating, the larvae chew indentations into the wood of the frame ends, tops, and bottoms, and the sides of the brood supers, and they often move up into the honey supers to spin their cocoons. You will certainly know that you had wax moths if you find their devastation in a hive that has died out or has become hopelessly queenless in the warm weather, usually after the swarm season has passed. To prevent infestation you must check the brood chamber at intervals. Be certain that there is a queen in the hive at all times, particularly after the swarm season has passed, just in case one of your hives swarmed and you did not realize it.

CONTROLLING WAX MOTH ONCE INFESTATION HAS SET IN

If you have only one hive, treat it the same way you would treat laying worker hives. Place each brood super, bees and all (if there are any), in a plastic bag with about four to five tablespoons of PDB moth crystals. If the infestation is very heavy, do not put the brood chambers in plastic bags, but stack them on a piece of plywood or on concrete.

Stack the supers, one on top of another, with two tablespoons of PDB crystals on a small sheet of paper between each super; then tape the supers together. Place several thicknesses of newspaper over the uppermost brood super, and cover this with the inner cover and hive top.

As the larvae begin to pupate and spin their cocoons, they can chew holes through a plastic garbage bag. The deadly PDB gas escapes, allowing the moths to continue to destroy the combs. *To repeat:* If infestation is heavy, do not place the brood supers in plastic bags, but proceed as directed in the previous paragraph.

If you have two or more hives and you find a hive with the first signs of wax moth infestation, unite the infested brood chambers, without newspaper, over a queen excluder and below any honey supers. If you have several hives, unite one brood chamber on each separate hive. If the wax moth hive had honey supers, get them on a strong hive as soon as possible, or if they contain no honey, store them away in plastic bags with PDB. If the infestation is bad, proceed as directed earlier for one hive, or strip out the messy comb remnants and burn or bury them, or somehow destroy the larvae and moths as quickly as possible.

If available, a good-sized swarm can be dumped into a hive with the first signs of wax moth infestation. The swarm will clean out the hive in almost no time. If you have time, watch at the entrance occasionally for the first day or so. You will see the larger moth larvae being pulled from the hive and carried away or, if they are too large to carry, towed away from the hive area.

UNIT 28

Removing Bees From Houses, Trees, Etc.

Removing bees from houses, hollow trees, signs, etc., is a costly exercise in terms of time and trouble. Most beekeepers will try it once before they learn it is not that much fun and is a difficult means of starting a hive of bees.

Do not bother with hives in structures that cannot be opened to view the combs for removal. Most people do not want their house's siding removed or stucco torn open or a tree cut down to remove the bees, but they want you to get the bees. Tell these people they cannot pay you enough money to do that sort of thing. But you will not believe me until you try it once and find that you do not like it or that you love it. If you are going to take bees out of a house, do it for the fun of it; otherwise, charge a fair amount for doing it. These strangely located hives in structures that cannot be moved or torn open to expose the comb and bees are referred to in this unit as *odd hives*.

Funneling or transferring bees from an odd hive works as follows. The bees are routed from the odd hive through dryer hose, then through a bee escape or through a cone of screen into the proper hive. Each day all the young bees that come out to make their first orientation flights will be funneled off. If there are brood and bees in the proper hive, the young bees will join up with that unit, because they will not be able to get back into the odd hive. If there is nothing but foundation in the proper hive, the young bees will have no reason to stay. After a few days, you will find that there are very few bees in the hive; they just seem to disappear. Funneling must continue for a period of four to six weeks, depending on the strength of the odd hive and the stores of pollen and nectar/honey in it.

In order to funnel bees from an odd hive, you must be able to provide *all* the following items. If any one of these items cannot be provided, the funneling will not work.

1. You must have access to the odd hive and a way to get the proper hive within a few feet of the entrance to the odd hive.
2. There must be only one entrance to the odd hive. If there are many holes available for the bees in the odd hive to use, all must be closed except the one from which funneling will be done. In some houses, there may be dozens of places the bees can exit when the original exit is covered. When the exit to be funneled is darkened, the bees will go to another exit that shows light, and they will not go down the funnel as you had planned. So, for the first few days you must watch that the bees in the odd hive have not begun to use an alternate exit.
3. You must be able to maintain the funnel system in place for a period of four to six weeks. The queen will continue to lay for a time after the funnel is fastened in place, and it takes three weeks for the metamorphic cycle from egg to adult in *Apis mellifera*. The new young bees do not leave the hive for the first five to seven days after emerging.
4. You must have frames of brood and bees from which the funneled bees can make their own queen, or a nucleus hive, a swarm, or a weak hive must be available in the proper hive.
5. When the bees are funneled out into the proper hive, there must be no way for them to return to the odd hive.

ACCESS TO THE HIVE

You must move the proper hive near the entrance to the odd hive in order to funnel the bees, using the shortest funnel possible. If the hive is up on the side of a house, it is sometimes possible to make a platform between a rung of a tilted ladder and a two-by-four attached to the side of the house. If the odd hive is in a tree, it may be possible to support the proper hive on a platform held in place by a pair of ropes hung from one of the stronger limbs. Use your ingenuity to make a place for the proper hive near the odd hive.

A funnel for transferring the bees from the odd hive to the proper hive can be a cone made of screen, or a tube, such as the flexible dryer hose, that can be bent around to the proper hive. A cone of screen can be made as long as necessary, within reason, if you have a big enough piece of screen. The screen is formed in the shape of a cone, with one end having an opening about ¼ inch in diameter and the other end having an opening large enough to cover the exit of the odd hive. The cone probably will have to be roughly sewn together to maintain its shape. It should be fastened to the exit of the odd hive with tacks or staples and the small end of the funnel should be placed either a few inches back into the proper hive entrance or a few inches into the oval hole of the inner cover. You must be sure that the small end of the cone in the hive entrance does not lie on the bottom board but is held up off the surface about ¼ inch so the bees cannot find their way back into the cone. This is also true if the small end of the cone is placed through the oval hole in the inner cover; make sure it does not touch or lie on anything.

A tube of flexible dryer hose is very convenient to use because it can bend around things and end up covering the oval hole in the inner cover. The flexible tube should be attached to the exit of the odd hive. In some cases, a special fitting must be made to hold the tubing at the exit or to cover an opening larger than the flexible tube. Use your ingenuity for this sort of thing. If the opening is small enough to be covered by the tubing, it can be fastened with tacks, staples, masking tape, etc. The opposite end of the tubing is placed over the oval hole in the inner cover *only after the bee escape has been put in place.* The tubing should be fastened to the inner cover, and the system then will be ready to go to work.

If you are going to funnel the odd hive through a hive containing bees with a queen, use an empty dummy hive to get the funnel ready. Then bring in the weak hive or the hive in which the nucleus brood frames and queen have been placed. If you are going to bring in a few frames of brood and bees to let the bees make their own queen, get the setup ready, and bring in the brood and bees about twenty-four hours after the funneling from the odd hive begins.

Once funneling from the odd hive has begun, you must check the area around the odd hive exit over a period of several days to make sure the bees have not found another way out of the odd hive. If they have, plug those exits as you find them. A rag jammed into the odd hole exit is the best plug material. Once funneling has begun, the old field bees will exit through the funnel, and they will fly for nectar and pollen and return to the old exit. They will hang around that exit for the rest of the day, presuming they cannot get back through a faulty funnel connection. Do not worry about those field bees, for they will join the proper hive at night, but do expect them to be around the odd hive exit for several days.

ONCE FUNNELING HAS BEGUN

Once funneling from the odd hive into the proper hive has begun, it is just a matter of time until most of the young bees will have been removed from the odd hive. You should wait at least five to six weeks.

After five to six weeks, you can remove the funnel arrangement from the odd hive entrance and from the inner cover or proper hive entrance and place the hive top on the hive. The bees in the proper hive will discover the odd hive entrance, and finding no guard bees on duty, will begin to rob honey from the odd hive. This is what you want to have happen, particularly if the hive is in the side of a house. If the bees rob out the odd hive, there will be no honey to run into the house when the comb melts and the honey begins to run during hot weather.

Leave the hive in place for about one week after the funnel has been removed so the bees can clean the honey out of the odd hive. This will not only be good for the proper hive, but it could save a mess between the walls of a house. When the honey is gone, wax moths will destroy the comb, and there will be nothing left but the webbing of the wax moth. The entrance to the odd hive should be closed after the bees have cleaned out the honey, or another swarm will find that opening during swarm season next year and move in to start another odd hive.

When all the honey from the odd hive has been cleaned out, it is time to move the hive to its permanent home. Close the hive at dark with a roll of screen. If you have to bring the hive down from a tree, a roof, or the side of a house, you may want to wait for the full moon so you can see what you are doing without using lights. Move the hive to your truck or to the trunk or back seat of your car and take it home—if you live more than a mile away. If the hive is in your yard or nearby, move the hive about two miles away to a friend's yard, as previously arranged. Remove the entrance screen when the hive is safely in its new location.

If you moved the hive to a friend's yard, wait two weeks. Close the hive with the roll of screen after dark, and move the hive to your yard in its permanent location. Remove the entrance screen. During the balance of the season add a second brood chamber and honey supers as required.

UNIT 29

Removing Bees When the Comb Can Be Exposed

It is not all that difficult to get bees from structures if the siding or plasterboard can be removed to expose the comb. The comb can be cut down, the bees shaken into a proper hive, and the brood comb salvaged until the brood emerges.

You can have a similar adventure removing bees from other containers that can be brought home, where you can transfer the bees into a proper hive at your convenience. Anything that can be closed at dark and put in a truck is worth considering for the bees and brood it contains. (See Figure 29-1.)

The same procedure used for getting bees from structures is used for getting bees from wooden boxes, birdhouses, chests, trunks, kegs, wooden barrels, etc. The main difference is that you can take the small odd movable hive home, where the bees can orient to your apiary until you find the time to tear the container apart. You might hold one container until you get another, or have several and tear them apart one after the other. If you pick up the odd movable hives near the end or after the honey flow, you may not want to break into them until spring. Many books and beekeepers will tell you to save the brood in the combs removed from exposed structures and odd hives, by cutting it to fit and tying it into empty frames. That sounds like a lot of work, but the worst part is that you soon have to toss that brood comb out. It is the worst brood comb in the world, packed full of drone cells. After you tie it in the frame, the bees fasten and fix it so that it has even more drone cells. Here is a much easier way to save that brood.

1. Brush the bees from the odd hives' wild comb into a proper hive with frames of foundation or comb. (Comb is preferable if you have it.) You brush the bees from the wild comb to get the queen down into the proper hive brood chamber.

REMOVING BEES WHEN THE COMB CAN BE EXPOSED 273

Figure 29-1, Some Odd Movable Hives. Clockwise from the top left: Hives of bees in a 55-gallon oak barrel, in a 20-gallon antique oak butter churn, in an old-style nail keg, and on a shelf attached to a redwood fence. Keep your eye on the old-style nail keg, because the bees are going to be removed from that keg in Figure 29-2.

2. The wild comb, which is of no use whatsoever except for the brood (eggs, larvae and covered brood), will be placed in an empty super after the bees have been brushed off. The empty super will be setting on an inner cover.
3. When all the comb has been cut down or cut out of the odd hive, brushed free of bees, and stood on end in the empty super on the

inner cover, place a queen excluder on the proper hive, and set the inner cover with the super of wild comb on the queen excluder. The young nurse bees will hurry through the queen excluder, through the oval hole in the inner cover, and they will tend the brood until it emerges. The queen must stay below the queen excluder, the older bees will draw foundation into comb, and the queen will begin to lay eggs in the comb in the proper hive.
4. Three weeks after you place the super of wild comb on the hive, all of that brood will have emerged. You can remove the top, place a bee escape in the oval hole of the inner cover, and remove the super of wild comb twenty-four to forty-eight hours later. By this time, all the honey that was in the wild comb will have been moved to the lower brood chamber, and the wild comb can be melted down or tossed in the garbage.

Be aware that cutting the comb from walls, boxes, etc., can be a very messy job. When the comb is cut, the honey will drip and get on the bees, which will make brushing the bees more difficult. The job should be done early in the day to give the bees that are covered with honey a chance to clean themselves off or have the other bees clean them off before dark and cold does them in.

Before you tackle one of these jobs, get a pair of plastic or rubber gloves, either to use or to put over your nice leather gloves, because this can be a messy job. You will want a bucket of water also. When things get too sticky, you can wash off your gloves and the bee brush.

The most important step in transferring bees and comb from any odd wild hive is to smoke it heavily, wait about five minutes, smoke it again heavily, wait about five minutes, smoke it again heavily, and wait another five minutes. While you are smoking and waiting, you can be gently tearing the siding from the house or gently tearing the movable hive container apart. Very seldom will you have the slightest trouble if the hive is smoked with a lot of *cool* smoke that is given time to take effect. If the portable odd hive can be inverted and the combs cut from below, you have about an 80 percent chance of getting the queen unharmed into the proper hive.

TRANSFERRING BEES AND BROOD FROM ODD HIVES

Two methods of transferring bees are described in this section. Method A deals with removing bees from structures such as walls of houses, sheds, and garages; Method B deals with transferring bees from odd movable hives such as birdhouses, boxes, kegs, chests, etc. Follow the steps below until the method callout sends you to Method A or Method B.

Equipment needed:
- [ ] Hive tool
- [ ] Smoker or hive bomb
- [ ] Bee brush
- [ ] Bucket of water
- [ ] Rubber gloves
- [ ] Hammer, crow bar, utility knife
- [ ] Ten-frame hive with top and bottom, and ten frames of foundation/comb
- [ ] Inner cover
- [ ] Queen excluder
- [ ] Empty super
- [ ] Dust pan or flat cardboard

After three weeks:
- [ ] Bee escape

Step-by-step directions:
- [ ] Set up the hive, and staple the hive bottom to the brood chamber. Load up everything you need, and carry it to the wild odd hive; set it up as follows.

  Set the hive so that as you remove the comb from the odd hive you can reach over and brush the bees from the comb into the hive (you will remove about three frames and stand them in a safe place).

  Set the inner cover on the ground nearby, and set the empty super on the inner cover. (This should be in a handy spot, because you are going to stand the wild combs that have been brushed free of bees in the empty super.) Stand the queen excluder in a handy place nearby. Fill the bucket about two-thirds full of water, and put it in a convenient spot; you will want to wash the honey from your gloves and brush fairly often. Have the hammer, crowbar, hive tool, dustpan, etc., handy.
- [x] **Dress properly,** if you have not had to do so by now.
- [x] **Light the smoker** or have hive bomb handy. Stoke the smoker with a full load of cool-burning materials, for you will be smoking the hive heavily.
- [ ] Smoke the odd hive entrance heavily, and wait about five minutes for the smoke to take effect. With the hive tool, hammer, or crowbar, begin to remove the siding from the structure, or remove one of the sides of the smaller movable odd hive. Work gently. When you have opened the odd hive a few inches and the bees have begun to come out to investigate, smoke the odd hive again heavily at the

entrance and in the small opening that you have just made. Wait another five minutes.

☐ Continue to remove the siding or the side of the small odd hive. It the bees appear in large numbers as you proceed, smoke them again heavily and wait a few minutes. As soon as the first strip of siding is off or the side is off the small odd hive and the comb is in view, smoke the comb heavily and wait about five minutes. As you are waiting, plan how you can best get the comb out. By this time the hive should be pretty well smoked, the bees well gorged and disoriented, and you can proceed without too much trouble. When you have removed enough siding or sides of the small odd hive, smoke once again between the combs in the exposed area. If you are transferring bees from exposed comb in the wall of a house, shed, or garage, follow the directions in Method A. If you are transferring bees from a smaller movable odd hive, go to Method B.

A. Transferring bees and comb from a wall in a structure.

☐ Now that the comb is exposed, brush the bees from the face of the comb onto a dustpan or piece of cardboard, and dump them into the hive (from which you will have removed about three frames that are standing in a safe place nearby).

☐ If the comb is quite large, it will be necessary to cut the comb to a size of nine inches deep and fifteen inches long to stand in the empty super that is on the inner cover. When the comb is cut, let it settle into your glove, move it to the hive, and brush the bees from the back side of the comb. Move to the empty super and lean the comb at a slight angle against the side of the super. You may find that smaller pieces are easier to handle and do not break apart so much, particularly if they are heavy with honey. If there is much honey in the comb, cut the brood comb away, brush the bees off, and place the comb in the super. Then cut the honeycomb out, brush off the bees, and throw it in a bucket for your own use. You should place some honey in the super with the brood, however.

☐ If your gloves and brush are sticky with honey, wash them off in the bucket of water, shake the brush free of water, and continue.

☐ When the second piece of comb is brushed free of bees and placed in the super next to the first piece, place a bit of comb between the two combs so they will be held about ½ inch apart at the top. You may need two or three pieces on longer combs. This will create space between the combs so that later the nurse bees will be able to tend the brood in the combs.

- ☐ Each time a new comb is exposed to view, wash the brush and shake it, then brush the bees from the face of that comb into the dustpan or piece of cardboard, and dump the bees into the hive. Continue to cut the combs, brush the bees from the back side, and place the comb in the super with the other brood combs, propping them apart with bits of comb until all combs have been removed from the wall.
- ☐ Replace any frames removed earlier from the brood super and place the queen excluder on the brood chamber. Lift the inner cover and the super containing the brood, and place it on the queen excluder. Place the cover on the hive. Before leaving the area, scrape as much of the wax from the wall cavity as possible.
- ☐ You can now leave the area and return about dark or sometime after dark. When you return, check the wall cavity for a cluster of bees where the hive originally was. It is hoped that the bees that are not too covered with honey to get cleaned up will have discovered the hive and joined up. If there are bees in the wall cavity, brush them onto the dustpan or cardboard, and dump them on a ramp in front of the hive. Do not dump them in among the brood comb, for the queen may be in that cluster. If there is quite a cluster of bees in the wall cavity and you dump them in front of the hive, you may want to wait until the following night to close up the hive, staple it together, and move it as suggested in the following step.
- ☐ If most of the bees are in the hive, close the entrance with a roll of screen, staple the super containing brood to the brood super on the hive, and move it to your place, if it is at least one mile away. Otherwise, move it to a friend's place about two miles away. Leave it for two weeks, then close the hive after dark and move it home.

Skip to Step C and proceed with the remaining steps for transferring bees from a structure to a proper hive.

B. Transferring bees from an odd movable hive. (See Figure 29-2.)
- ☐ Invert the odd hive and move it so that the proper hive can be set in the exact location previously occupied by the odd hive. Any bees that have oriented to the location when flying from the odd hive will fly to the proper hive and may be saved from a honey bath.
- ☐ When the odd hive is inverted, the combs will be pointing up rather than hanging down; you will be slicing the comb from below rather than from above. This arrangement, if you can manage it, leaves most of the honey mess at the bottom, and the comb can be moved and the bees brushed without getting honey drippings over the comb and the bees.

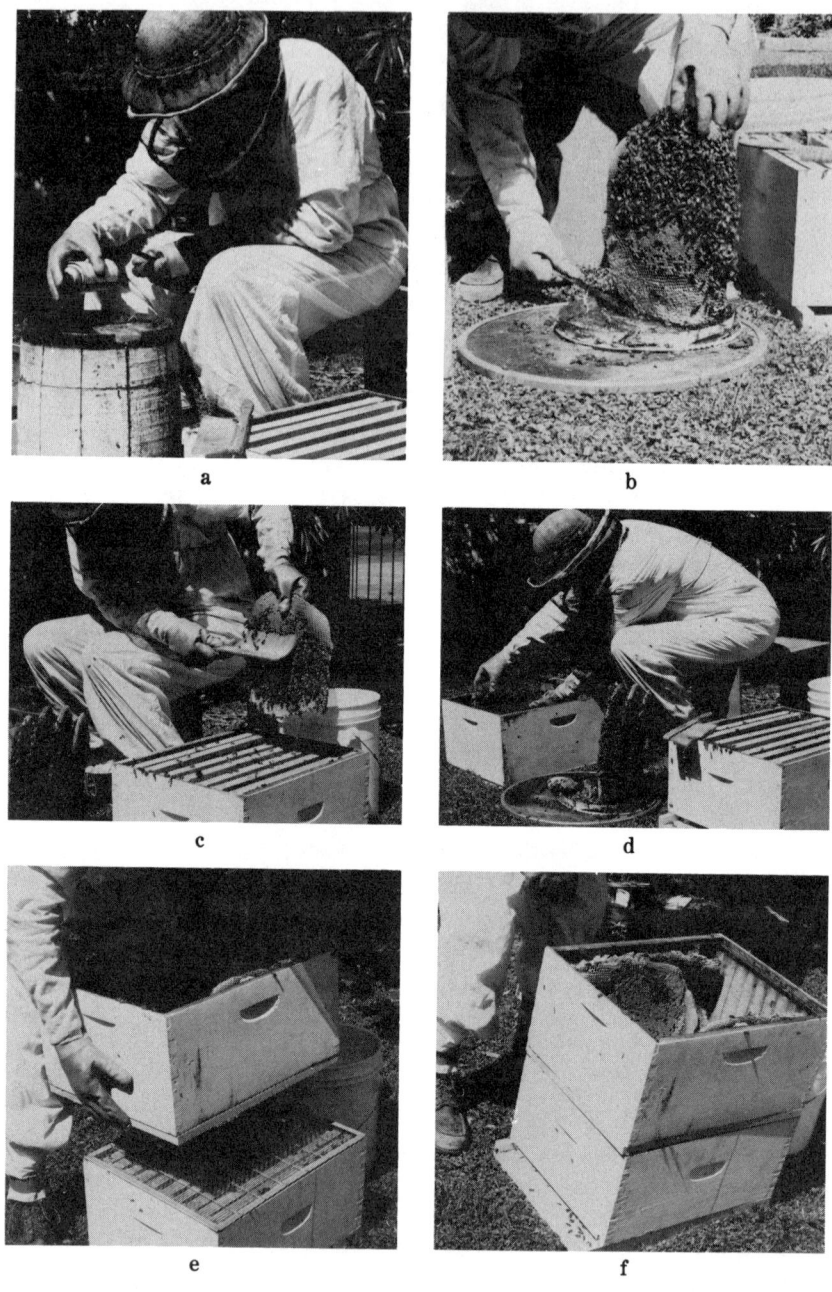

a b c d e f

☐ Remove three frames from the hive, and stand them nearby in a safe spot. Set the inner cover nearby, and set the empty super on the inner cover. You are now ready to cut away the first comb by slicing along the comb where it is attached. If the comb is quite large, you may have to cut it to a size that will fit in the empty super, nine inches by fifteen inches.

☐ Lift the comb, move to the hive, and brush the bees from both sides of the comb. Move to the empty super and lean the comb at a slight angle against the side of the super.

◀ **Figure 29-2, Removing Bees from an Odd Movable Hive.** Six photos showing how the bees were removed from the keg in Figure 29-1. a. After smoking the keg heavily several times, the keg has been inverted and wire bands removed, and the bottom of the keg is being removed. b. The keg staves have been removed, and a comb is being cut from what used to be the top of the keg. c. The comb just cut from the odd hive is being brushed free of bees and put into the proper hive placed conveniently nearby while the author is comfortably seated on an empty super. Notice the wild combs standing upside down on the odd hive lid. d. To save the brood, the wild comb of brood is being placed in the empty super that is on an inner cover in a convenient location. Notice the bucket of water nearby. e. After all the combs have been cut, brushed free of bees, and placed in the empty super, a queen excluder is placed over the proper hive into which the bees and queen have been brushed. The inner cover and empty super that is now full of wild combs of brood are lifted up onto the queen excluder. f. The hive is now ready to be closed up. In three weeks, all the brood in the upper super will have emerged, and the wild comb can be melted down or tossed in the garbage. Notice the capped brood on the comb in the upper super. Very soon after setting the upper super of wild brood comb on the hive, the nurse bees will be through the queen excluder, tending to the brood.

☐ Anytime your gloves and bee brush become sticky, wash them off in the bucket of water, and shake the brush free of water before continuing.

☐ Slice the next comb free. Pick it up and brush the bees from both sides into the hive. Move to the empty super and lean the comb against the comb in the super. Place a bit of comb between the two combs so that they stand about ½ inch from one another. This makes it possible later for the nurse bees to tend the brood in the combs.

☐ Continue removing the combs from the odd hive, brushing the bees, and setting the combs in the empty super until all combs have been removed from the odd hive.

☐ Replace the frames removed from the hive earlier. Place the queen excluder on the hive. Lift the inner cover and the super containing the brood comb onto the queen excluder. The nurse bees will hurry through the queen excluder and the oval hole in the inner cover and continue to tend the brood. The queen, who has been brushed into the brood chamber of the hive, will not be able to return to the wild comb above. (See Figure 29-2.)

☐ Knock any bees on the odd hive parts off in front of the hive, and remove the remnants of the odd hive from the area. Continue with the following steps.

C. After the bees have been transferred.

☐ *Wait two weeks.*
■ **Dress properly.**
■ **Light the smoker** or have hive bomb handy.

■ **Enter the hive.** Invert the hive top on the ground nearby. Smoke between the inner cover and the queen excluder. Wait one minute, and set the inner cover and the super containing the wild brood comb onto the inverted hive top. Remove the queen excluder.

☐ Check the brood chamber of the hive to see that comb is being drawn and that the queen has begun to lay. It was mentioned earlier that comb is preferable to foundation, but the bees should have begun to draw comb if the honey flow was not over. If you find eggs and young larvae, you know that you transferred the queen to the brood chamber.

If the queen is in the brood chamber of the hive, replace the queen excluder, set the inner cover and the super of wild comb on the hive, and close the hive. Wait one more week. Enter the hive, place the bee escape in the oval hole of the inner cover, and wait two more days. Remove the inner cover and super of wild comb. Melt down the wild comb for wax, or toss it out in the garbage.

☐ If you find no eggs or young larvae in the brood chamber of the hive, you will know that you did not transfer the queen or that she was lost in the transfer. In this case (1) you will find the queen in the super of wild brood or (2) if she was lost, there may be queen cells in the wild brood.

1. If you find eggs and larvae in the wild comb, you will have to smoke that super of comb heavily, wait five minutes, remove the wild comb from the super piece by piece, and brush the bees and queen down into the brood chamber of the hive. Set the brushed comb aside or in another empty super that is on an inner cover. Replace any frames removed from the hive, replace the queen excluder, and set the inner cover and super of wild comb on the queen excluder. Wait two weeks, and check that the queen is in the brood chamber of the hive. If so, wait one more week and remove the super of wild comb using the bee escape.

2. If you find queen cells in the the wild comb, you will know that the queen has been lost in the transfer and that the bees are making a new queen; in fact, after two weeks, the new virgin queen probably will have already emerged. About all you can do is:

   a. Follow all the steps in item 1 and get the virgin queen into the lower brood chamber. Check in three weeks to make sure that the queen has begun to lay.

   b. Follow all the steps in item 1. Order a mated queen; when she arrives, de-queen the hive and introduce the new mated queen. (See Unit 13.)

Caution: In the North, it is best if you pick up portable odd hives after the honey flow is more than half over to bring them home and let them set until spring. If the bees live through the winter, transfer them in the spring. Otherwise, transfer the bees knowing that you will probably have to unite the hive in the fall.

SOME TIPS ON STRANGE ODD HIVES

1. An oak barrel with bees coming and going from the bung hole.

   To get into a barrel such as this for transferring the bees and brood, smoke heavily through the bung hole. While waiting five minutes, mark with chalk the center of the barrel where you are going to cut the barrel in half with a power handsaw. Turn the barrel on its side, smoke heavily, and cut the barrel in two as you roll it along on the ground.

2. A wooden keg.

   Smoke under and into the keg, and wait five minutes. Invert the keg and smoke between the combs. Cut the wires holding the keg staves together, and begin to remove the staves, smoking between the combs as necessary. This method will also work with oak barrel halves if you do not want them for garden use. (See Figure 29-2).

3. A ceramic pot.

   Smoke under and into the pot, and wait about five minutes. Invert the pot and smoke between the combs. With a hammer, crack the ceramic and remove the pieces. Continue to crack and remove the pieces until you can get to the comb to cut it loose. As you proceed, smoke between the combs as necessary.

4. A metal tub or other metal container.

   If you cannot cut the metal container with tin snips (after smoking it heavily), then you will have to scoop out comb and honey along one side until you can make an opening from which you can remove combs or sections of combs. This scooping is quite messy, but sometimes there is no other way.

Before and after the honey flow, if you are going to tear into an odd movable hive that will be a sticky mess, move it to an area away from the apiary to perform the transfer operation, or you may start robbing in the apiary. If the sticky mess is some distance from any hive, the chances of robbing will be reduced. After the transfer is complete, move the hive back into the original odd hive location, and close the entrance to about one inch. If transfer is made after the honey flow, you might also consider making the transfer as late in the day as possible to prevent robbing.

UNIT 30

Rearing a Few Queens for Your Own Use

If you are going to rear queens, you ought to have a use for them. Use them to re-queen your own hives or to start new hives from divides, or keep several queens on hand just for replacements, if needed. In most cases, it is best to keep the new queens in small nucleus hives until they have had a chance to mate and have begun to lay. If you re-queen with a virgin queen, it may be about five weeks before the first new bee emerges in the hive. There is also the risk that a queen will not return from her mating flight, so it is best not to subject a productive hive to a virgin queen. Get the queens started in small units so that you will have a mated queen with which to re-queen.

If you remove the queen from any hive for several days, the bees will begin to make a number of queen cells. As long as the old queen was laying at the time of her removal or loss, the hive can and does re-queen itself. Just keep in mind that a new virgin queen will not always return from her mating flight.

Within several hours of the removal or loss of a queen in the hive, the bees begin to select one or many worker larvae (from eighteen to thirty-six hours old) as replacement queens. Many are selected, but in the end, only one will serve as the new mother of the hive. A number of queen cells are developed to make sure that at least one will survive. In very small units, such as observation hives, the bees usually will start at least four queen cells.

Replacement queen cells are actually emergency queen cells. If the colony is to survive when the old queen disappears, it is very essential that a new queen be reared without delay. Nature has set it up so that the metamorphic cycle of the honeybee queen is completed in only sixteen days.

The larvae selected to become queens in replacement situations generally are found on the face of the brood comb and, less often, at the lower part of the frame, closer to the bottom bar (where swarm queen cells usually are found). In emergency situations, selection must be made from the cells in which larvae of the proper age are found. The larvae selected are fed copious amounts of royal jelly and literally are floated out of the brood cells. As the queen larva grows, a special cell, called a queen cell, is built around it. A queen cell is built over an area of about three brood cells. It hangs vertically from the brood comb, whereas the workers and drones are reared in the usual fore and aft cells in the brood comb. The bees continue to feed the selected larva royal jelly in large quantities during the six days of larval life. Queens are developed from the larvae of fertile eggs (worker larvae) if those larvae are fed royal jelly for the full larval stage.

When the larval stage ends, the queen cell is capped and the prepupae and pupae stages begin; they last for seven days in the case of a queen honeybee. Sixteen days from the time the egg was laid, an adult queen emerges. When the first queen emerges, she probably locates the other queens by pipping. If she gets a response from other queens in the cells, she finds them, cuts a hole in each cell and stings each occupant without ever having to meet in mortal combat. If two or more queens emerge simultaneously, the queens must meet in mortal combat. At some point, the bees take a hand in the competition and begin to tear down all the younger queen cells so that only one queen will remain in the hive.

When rearing queens, it is necessary to remove the queen cells about two days before the first queen (from the oldest larva started) emerges. Give them individually to queenless hives, or enclose the queen cells in screen tents where the new queen and the bees cannot get to the queen cells to destroy them. In most cases, failure to remove or protect the queen cells will result in all but one queen being destroyed.

Before reading the following suggestions on raising a few queens for your own use, be aware that several books on the market cover the rearing of queens. They explain queen rearing for the hobbyist beekeeper who wants only a few queens and for the operator who wants to rear queen bees as a commercial business.

RAISING A FEW QUEENS

Pick a hive that has a record of gentleness and high productivity from which to produce your queens. This is a form of selective breeding, even though it is only on the queen's side. Unless you inseminate the queen artificially, you cannot control the selective mating of the queen by the drones in the air. In this competition, only the strongest and swiftest mate with the queen, and you have no control over whether the drones come from a genteel background or from hard-working stock.

Queens from swarm cells are probably the best queens because the queen larvae usually are larger and more well fed. Queens reared from the egg in small hives and at times of the year when nectar and pollen are not available in abundance may be smaller and less prolific. Most queens reared in a strong hive during the period that nectar and pollen are available in abundance in the field will probably be good queens.

QUEENS FROM SWARM QUEEN CELLS

If you want to use swarm queen cells from a gentle and productive hive, you must keep watch on the hive. When it begins to prepare queen cells for swarming, check the hive at one- or two-day intervals until the queen cells begin to be capped. As the cells are capped, you can push a wire screen tent (about one inch high, one inch wide, and three inches long) over the queen cell. Push in as many tents as you need queens. As more queen cells are capped, you can tent more of them. When the first queen that is *not* tented emerges, she cannot sting the tented queen cells. Seven days after the queen cell is capped, the new queen emerges. As you check the frames and the tented queen cells, you will begin to see the young queens moving in the screen tent. You can lift one side of the tent and get the young queen into a queen catcher or small jar. Young virgin queens cannot be kept caged for long. If the young queen does not mate within the first few weeks of her life, she will not mate and will become a drone-laying queen. As a general rule, a queen that has emerged from a queen cell may not be accepted as readily as a queen that is given to a queenless hive in the queen cell; that is, before the queen emerges. If she emerges from a queen cell in the hive, her acceptance is almost certain. It may be possible to introduce a very young virgin queen directly into a queenless hive and have her be accepted, but it is best to introduce the young virgin in a cage for several days.

The reason to tent the queen cells is that you will not know when the first queen will emerge, because they were started over a period of about a week. When the first queen emerges, all the other queen cells soon will be destroyed unless they are protected.

GENERATING EMERGENCY (REPLACEMENT) QUEEN CELLS

In order to get queen cells of nearly the same age—if you do not want more than about eight queens—you can proceed as follows.

Remove the queen from one of your strong, gentle, productive hives.

If you need a queen for immediate re-queening, use this queen for re-queening another hive; otherwise, introduce the queen into a nucleus hive.

Within about twenty-four hours, the queenless hive will have selected young worker larvae less than two days old and begun to generate new queens. Because these are emergency queen cells, most of the queens will be started within two days, so they will be nearly the same age. Queen cells may be started on one or several brood frames.

You know that a queen's metamorphic cycle is sixteen days. The emergency queens are starting with larvae about two days old, which means that the oldest queen cell will release its adult virgin queen in another eleven days (three days for the egg to hatch and two days of the larval stage).

Wait ten days *only*, and then remove the queen cells by cutting them out and pushing them into the brood combs of the queenless hives or the nucleus hives you will set up for queen mating. If you wait more than ten days and the first queen emerges from her cell, it will be only a matter of hours until all the other queen cells are destroyed.

To remove the queen cells from the frame or frames, cut deeply into the comb in a cone shape around the top of the queen cell. Be very careful not to damage the elongated portion of the queen cell, or the hive to which it is given may not accept it and may tear it down. Hold the queen cell at the top or attachment area, where it is the strongest. While holding the queen cell at the top, you can push it into the comb of the nucleus hive or any other hive to which the queen cell will be given.

In most cases, the queen emerging from the queen cell in a hive will be accepted by the bees in the hive. Within three weeks, inspect the brood chamber to which the queen cell was given to make sure the queen has mated and begun to lay.

Most beekeepers with a few hives usually can maintain an observation hive as a source of a good-quality replacement queen, if one is needed. If the queen is removed, the observation hive will generate a new queen, and you can watch the process. Like the observation hive, nucleus hives will generate new queens when you remove their queens to re-queen or add a queen to a queenless hive. You may want to keep several nucleus hives for replacement queens, in case you need them.

Caution: It is most important that there be drones available to mate with the queen when she makes her mating flight or flights. If too few drones are available, the queen may mate poorly, soon run out of sperm, and become a drone-laying queen. Remember that a drone must be about ten days old before it is mature enough to mate with a virgin queen.

Do not start too early in the spring to rear queens or make divides in which the bees will rear their own queens. Drones may not be available.

Do not attempt to rear queens too late in the season. Do not attempt to rear queens during the winter, because drones are driven from their hives in the fall, and there may be too few drones to accomplish a good mating for the queen.

UNIT 31

Re-Queening a Hive of Mean Bees

We can be thankful that meanness or overaggressiveness in bees is a hereditary trait. If it had not been so in ages past, it is possible there would be no bees for us to enjoy today. Humans and animals long ago would have plundered all the hives faster than the bees could have reproduced. Meanness probably begins because hives swarm and a virgin queen mates with drones from a more aggressive hive. Meanness, presumably a dominant trait, continues and no doubt increases each year, so it is wise to re-queen the hive occasionally with a new, gentle, mated queen. Remember that the queen and drones are responsible for the gene traits of the bees in the hive. Most queens produced by queen breeders and producers today are relatively gentle.

Responsible beekeepers keep their hives as gentle as possible to avoid difficulties and confrontations in the neighborhood. The author feels strongly that if a beekeeper does not do everything possible to maintain gentle bees and to tend those bees in a proper manner, his beekeeping operation should be shut down. This remark does not imply that any complaint is reason to move bees from the neighborhood or that the beekeeper is not tending the bees properly. The fear of bees often causes uninformed people to complain when in fact the bees cause no trouble whatsoever.

As you can imagine, mean bees are not much fun to work and often can keep beekeepers and their families from using the area in which the bees are located. It is not much fun, either, to tear into a mean hive to de-queen it in order to re-queen with the new, gentle mated queen. Generally, the older bees do the attacking and stinging. Even mean hives are easier to work during the honey flow and in the middle of the day. When working mean hives, use lots of smoke, and give the smoke more time to take effect.

DE-QUEENING AND RE-QUEENING A MEAN HIVE

When you have the new mated queen in hand, proceed as follows.

☐ If there are honey supers on the hive, remove them before the new queen arrives so that you will have only the brood chambers to move and search.

☐ After dark, ■ **Dress properly.**

■ **Light the smoker** or have hive bomb handy. Smoke the entrance of the hive to drive in any bees hanging out on the entrance. Wait a few minutes, and then insert a roll of screen into the hive entrance. After the hive entrance has been closed, staple the brood chambers together, and staple the bottom board to the lower brood chamber.

☐ Move the hive some distance from its present location—at least ten feet, but more if possible. Pull the screen and leave the area.

☐ Set up a temporary dummy hive on the exact location from which you moved the mean hive. This can be a piece of plywood on which a super with frames of comb or foundation is set and propped up so the returning bees can enter and cluster the next day. Place another piece of plywood on top for a cover. Have this unit ready before dark.

At first light, the old bees will begin to leave the relocated hive; when they return from the field, they will go to the old location. During the day, most of the old bees will leave the hive and return to the dummy hive on the old location.

☐ About 3:00 to 4:00 P.M., ■ **Dress properly.**

■ **Light the smoker** or have the hive bomb handy. Gently remove the hive staples between the upper and lower brood chambers. Smoke the mean hive heavily at the entrance, under the cover, over the upper brood chamber, and between the upper and lower brood chambers. Wait about five minutes. You may want to smoke the hive heavily a second time, depending on how mean the hive has been in the past. Wait about five minutes. Remove the hive cover and invert it on the ground nearby.

☐ Set the upper brood chamber onto the inverted hive top, and follow the steps in Unit 13, "Searching the Frames for the Queen." If you fail to find the queen by this method, then try the steps for "Dumping the Bees to Find the Queen" in that same unit.

Use lots of smoke as you remove the frames in the previous steps. You have one advantage here: All the older bees that fly out will return to the old location and should not be hanging around. When you find the old queen, capture her in a queen catcher or small jar, and remove her from the area.

REQUEENING A HIVE OF MEAN BEES

- ☐ Re-queen the hive as directed in Unit 13. Close the hive and leave the area.
- ☐ After dark, ■ **Dress properly.**
- ■ **Light the smoker** or have hive bomb handy.

  Go to the dummy hive. Pull that hive about five feet from its present position. Go to the re-queened mean hive. Close the entrance with the roll of screen. Re-staple the hive before you move it. Return the re-queened mean hive to its exact original position. Remove the screen and leave the area.

  At first light, the field bees in the dummy hive will begin to return to the original hive. Leave the dummy hive until dark, when you can dismantle it and store it away.

- ☐ *Wait four days.*
- ■ **Dress properly.**
- ■ **Light the smoker** or have the hive bomb handy.
- ■ **Enter the hive** by smoking the hive heavily at the entrance and over the upper brood chamber. Wait about five minutes.
- ☐ Check to see that the queen has been released; if not, release her manually. See Unit 13, "After Four Days."
- ☐ *Wait seven days.*
- ■ **Dress properly.**
- ■ **Light the smoker** or have hive bomb handy.
- ■ **Enter the hive** by smoking heavily at the entrance and over the upper brood chamber. Check to see if the queen has begun to lay. Smoke heavily between the first few frames removed. Follow the steps in Unit 13, "After Seven Days."

In forty-five to sixty days, all the bees in the hive will be the offspring of your new, gentle queen, and the hive should again be a pleasure to work.

UNIT 32

Robbing

IS IT ORIENTATION, SWARMING, OR ROBBING?

As you observe your hive, you will most often see the bee's normal comings and goings in the gathering of nectar and pollen. At certain times of the year you will see more activity than at other times, which you will learn to expect as you gain more experience. There will be times at the hive when you will see things happening that you may wonder about until you have experienced them a few times.

One of the activities that you will see most often, almost every day at almost the same time, is the orientation flight of the young (house) bees. Orientation occurs long before your hive population becomes so strong that there are thousands of bees in the air during the orientation period. As the population of the hive increases in the spring, orientation flights of more and more young bees (flying from the hive to eliminate body waste, exercise their wings, and learn the location of the hive) can make the less experienced beekeeper think that the hive is swarming or being robbed by other bees. Orientation usually occurs in the afternoon, lasts about thirty to forty-five minutes, and then the hive activity returns to normal.

Other activities that you may see around the entrance of the hive, but not nearly so often, could be swarming or robbing of the hive by another hive or by several hives.

HOW TO IDENTIFY ORIENTATION FLIGHTS

Two things can tell inexperienced beekeepers that what they are seeing are orientation flights.

1. Walk up beside the hive and look at the bees flying nearest the hive. You will see some bees that are flitting back and forth, up and down, all facing the hive and not more than a foot or two from the hive.

These are the very young bees (about seven days old) making their very first flights, and they don't get very far away from the hive. Depending on their ages, the older young bees fly farther away in wide circles, and even older bees may strike out in ever-widening circles and go some distance from the hive; they will go a bit farther away every day. In this way, they become familiar with the area. When they are about twenty days old and become field bees, they can navigate to and from the hive up to two miles in any direction.

2. Look at the landing platform of the hive and you will see a number of bees facing the hive entrance with their heads down, abdomens high, and wings fanning. These bees are scenting so that the very young bees can find their way back into the hive. The very young bees will fly from the hive about a foot and turn to look at the hive. They will fly up and down and from side to side, perhaps around the corner, and then back. After a while, they will land on the entrance, rest a bit, and then fly off and do the same thing again. The next day they will fly out two or three feet, and they may fly almost around the hive. Each day they go farther and farther from the hive. The orientation flights last from thirty to forty-five minutes, and then the hive settles down to normal activity. (See Figure 32-1.)

Mistaking Orientation Flights for Swarming

The inexperienced beekeeper may not recognize orientation and, seeing thousands of bees in front of the hive, may think the bees are swarming. If you have had a swarm come from your hive and later see orientation, you may think the bees are swarming again.

When the bees swarm from the hive, they leave filled up with honey. If you watch the front of the hive, you will see many of the bees crawling up the front of the hive before launching themselves into the air. This is because they are so laden with honey that they need some altitude so they won't drop onto the ground before they reach flying speed, just as it takes more runway for a heavily laden plane to lift off than it does for a light plane to lift off. Those bees that do not climb up the front of the hive come dashing out and seem to jump up into the air as they leave the hive; in other words, they seem to be make a running start and jump to get some altitude at takeoff.

Mistaking Orientation Flights for Robbing

All hives may have robber scouts. It is very easy to spot the occasional robber bee around the hive; in fact, their activities are a dead giveaway to the beekeeper and probably to the guard bees at the hive

Figure 32-1, Bees Scenting.
Above: Several bees are scenting on the hive entrance just moments after a swarm was top dumped among the frames. Left: Many bees are scenting after the bees were shaken from a frame of brood in front of a hive.

entrance. The occasional robber bee flits very nervously around the front and the corners of the hive, acting as though she is going to land. All of this nervous flight is usually within a foot or so of the hive, which can make you think that your hive is being robbed when all the bees in the orientation flights and the new bees are flitting nervously near and around the hive.

As you continue to watch the occasional robber bee, you will see a guard bee jump into the air and grab the robber bee. The two bees will fall to the ground, clinging together, and roll about like a marble gone mad. Soon they will break apart, and the guard bee will fly back to the entrance. Often the robber will take her leave but may go back to flitting nervously around the hive until she is grabbed again.

As a rule, if you do not see any bees scenting at the entrance when you see bees flitting around the front and sides of the hive, you are seeing a few robber bees that are around most hives when nectar is scarce. If the hive is relatively strong and the guard bees continue to grab and frighten away the robber bees, then nothing will happen.

If you find a number of pairs of bees clinging together and rolling about on the ground, you can assume that the hive entrance has been breached—the robber bee has gone home with a load of honey and passed the word. Now the reserves have been called out to continue the robbing, and the guard bees at the hive are engaged in fighting off the increased robber force. If you see more than two or three pairs of bees rolling about in front of the hive you can be quite sure robbing is going on.

ROBBING BY BEES

1. Leisurely robbing: Hive robbing is most often of the leisurely type. A robber bee finds a dead or weak hive—one that it can slip into—and then goes home with a load of honey. The word is passed, the bees from the hive are recruited, and they begin to carry the honey home. As a general rule, the hive that breaks into the weak or dead hive is the one that robs it out, for they do not allow any strangers in on the treasure.

    The following example demonstrates leisurely robbing. Few experienced beekeepers can say it has never happened to them.

    At one time or another you will have a weak hive, and as you observe it, you may see only a little activity at the entrance. Then one day you may begin to see a fair amount of activity and think that the hive surely has revived and is getting on fine. A few days later, there may be no activity at the hive, and that is when you may look in and discover that, though it may have a few bees and even a queen, the hive has no honey stores left. What will have happened is that the hive was robbed out so leisurely that there was no fighting between other strong hives. There was activity at the hive, but it was not frantic enough for you to suspect anything more than that the hive was doing better.

This sort of robbing in ages past was nature's way of clearing out the dead and dying hives. Now, unfortunately, it is a way of transmitting bee diseases. After experiencing this a time or two, you will begin to see the way it works and take steps to stop the robbing and salvage the honey and comb for restocking later. Do one thing if you suspect that this sort of robbing is going on: Watch for a while at the entrance to see if pollen is being brought into the hive. If the hive is active, you should see pollen coming into the hive to feed the brood, for there should be brood rearing if there is a goodly amount of activity. It is not very likely that bees are going to be bringing pollen into a hive they are robbing.

2. Frantic robbing: This type of robbing is caused by the foolishness of the beekeeper, either because of stupidity or inexperience. Beekeepers who toss sticky burr comb out of the hive or place it on the entrance for the bees to clean up or drip honey around the apiary in the vicinity of the hive are creating the setting for frantic and frenzied robbing of one or several hives in the apiary.

When nectar is scarce (before or after the honey flow), honey dribbled on, in front of, or around the hives will attract bees from all over the neighborhood. As the clean-up of the honey proceeds, bees arrive by the hundreds, and some begin to enter the hives and steal honey. Fighting will begin, more and more bees will be recruited from the robbing hives, the excitement will mount, and the fighting will increase. The noise of the frenzy can be heard 50 to 100 feet away. If one hive is closed off, the frenzied robbers will attempt to rob one or several other hives nearby.

There is probably no case of frenzied robbing of such proportions in the natural state that can compare with the magnitude of frenzied robbing in an apiary where honey is handled carelessly.

Precautions to Prevent Frantic Robbing

- Place sticky burr comb in a covered can or pan rather than tossing it out on the ground or placing it on the hive entrance.
- Take care that combs do not leak on or near the hives. If you can't find some way to contain the leaking, hose down the sticky mess before you leave the hive area.
- Do not leave a hive open for any length of time when there is no nectar flow. Cover the hive and any supers removed from the hive if you must move any frames or supers. If you must move frames or supers, try to do so as late in the day as possible so there won't be so much time for robbing to start before darkness falls and flying ends.

- Do not put sticky utensils or extracting equipment on or near any hives in the apiary during the day. Do not even put them out in the yard during the day. It is always possible that after the bees have cleaned up the mess, they will look for the nearest hive to enter, which could be accomplished easily because there are hundreds of bees ready to rob together.
- Make sure the top cover fits snuggly when you remove the inner cover to use it as a clearing board for removing supers of honey. If you see a large number of bees around the hive after you put the inner cover and bee escapes on, go have a look. Lots of bees will mean that somewhere there is an opening large enough for them to enter the super and rob out the honey and perhaps later start robbing in the apiary.
- Close down the entrance of any weak hives in the spring and fall. Reduce the entrances of weak hives when the nectar is not flowing. The smaller opening will allow a weaker hive to guard the opening more adequately.

What to Do If Robbing Starts

Prayers do not help when robbing starts in your apiary. You will have to help yourself somewhat by first closing the hive being robbed.

If robbing is just starting, close the hive to about a one-inch opening, and toss a handful of grass over the opening. If robbing is really going strong, close the entrance completely for an hour or two, and then open it to about a one- or two-bee-width opening (¼ inch), and cover the opening with grass. Leave the small opening for the balance of the day. After dark, enlarge the opening to about one inch, and again cover the opening with grass.

If robbing is really bad, you may want to close the affected hive or hives, set up a rotating water sprinkler nearby, and turn on the rain. The robbing bees should be slowed down rather quickly if they get soaked trying to get into the closed hive. If your hives are set up properly and tilt forward slightly, there is no fear of drowning the bees in the hive or the hives nearby.

When you are removing honey supers and robbing starts as a result of a nicked, broken, or warped super it is time to get out the masking tape to close the nicks or openings. For a warped top, put a flat rectangle of masonite or plywood under it to stop the robbing there. It will then be necessary to close down the hive entrance to about one inch and cover it with grass until the next afternoon.

Try to remember that if hundreds of bees in the area have been robbing a hive and you cut off their supply, they will be likely to go to the nearest entrance or entrances and attempt to break in en masse. The attacked hive may be taken so quickly that the coup is accomplished very easily. It is wise to close the entrances of hives near a hive being robbed so that they will not be overcome. Close the entrances to about two or three inches, and toss grass over the openings.

With normal care in the apiary, robbing will not be a great problem. Keep the entrances of weaker hives reduced longer in the spring, and reduce their entrances earlier in the fall, after the honey flow ends.

UNIT 33

The Observation Hive

If you want to gain an understanding of what really goes on inside the hive, an observation hive is most instructive. You will have hours of enjoyment watching the varied activities of the bees, both day and night. Most living things become distracted or frightened when a nose is pressed against the glass for a better view; bees do not care who is watching or from how close.

Observation hives can be obtained from bee supply houses, but unfortunately, they are rather expensive. The cost keeps many beekeepers from having the pleasure of watching the mysteries of the hive at close range. If cost is no object—and in this case, the expense might be worth it—or if you can put it on your birthday or Christmas list, get one of the bee supply shop observation hives. Otherwise, if you are handy and can get plans for an observation hive from a university extension service, county agricultural agent, or county agricultural commissioner's office, you can make an observation hive without too much difficulty.

SOME DON'TS REGARDING OBSERVATION HIVES

Do not expect to get anything from an observation hive but enjoyment. Do not spend a lot of money on an observation hive that is supposed to store honey for your use. If you want honey, get a standard, full-sized hive. If you want to watch the bees, get a simple observation hive. The number of bees that can be kept in an observation hive makes it difficult for them to store much honey except during the height of the honey flow. If there is a short break in the nectar flow, you will have to feed the bees a sugar solution. Observation hives can get so crowded that the bees may swarm during the honey flow. Later you will learn how to start the observation hive and how to keep it from swarming.

Observation hives can be kept through the winter in southern areas, but in the North they will not make it through the winter. In the North, start the observation hive each spring, and place the frames and bees in the big hive in the fall. In the South, the observation hive will have to be fed from November through February unless there is a winter flow in your area.

Do not build a big observation hive. There are plans available for an observation hive that holds four deep frames, one above the other. This observation hive is too big for most homes, and it takes two people to move it in or out when it requires attention. The observation hive that seems to work best for the home is one that holds a deep frame with a shallow or medium frame above—nothing much larger than two deep frames, one above the other. This observation hive is not too heavy to move in and out of the house for necessary manipulations. It can be hung from the wall with a swinging door hinge. The observation hive is connected to the outside with plastic tubing (the travel tube) through a window or wall, similar to the way a television cable comes into the home. Decide where the observation hive will be placed, or where it will hang, and have the tubing connected before you stock the hive with bees. When everything is ready and working to your satisfaction, you can stock the observation hive with bees as described later in this unit.

Do not make an observation hive that has several frames hanging side by side. This defeats the purpose of the observation hive; because the queen prefers the darker areas, you will seldom see the queen on the outside frames. If there are single frames, one above the other, the queen will be on one side or the other and you can watch her at any time of the day or night. Seeing the queen and the activities around her is one of the pleasures of the observation hive.

If you build your observation hive, you should make some provision for feeding the bees. Drill a tapered hole in the hive top into which you can insert the plastic spout from the top of a plastic honey bear. The tip of the spout should reach just a fraction of an inch below the underside of the top so the bees can sip the solution from the end of the spout. You can make another feeder arrangement by cutting a hole in the top of the observation hive just large enough to take the perforated lid of an entrance (or Boardman) feeder. Cover the hole on the underside of the top with screen so that the bees will not come out when you refill the feeder jar. The feeder jar can be a regular mason-type half-pint or pint jar, which looks nicer than a large quart jar on top of the observation hive.

You should provide a few vent holes on the ends of the observation hive. Some purchased hives have saw kerfs in the ends to permit ventilation. If you cannot make saw kerfs, drill a series of ⅛-inch holes (no larger) in the ends of the observation hive. The bees may close the vents with propolis, but they can be opened with the blade of a knife or with a nail pushed through the holes you have drilled. Venting becomes more important, particularly in the South during the winter, when moisture builds up inside on the glass and in the exit tube. If this begins to happen, unplug the propolis-covered vent holes.

STOCKING THE OBSERVATION HIVE

Two methods of stocking the observation hive are discussed here. One way to stock an observation hive is to borrow a frame of brood and bees from a hive and either let the bees make their own queen or add a new young mated queen at the time the observation hive is stocked with the frame of brood. A second way to stock the observation hive is to dump a small swarm (one to two pounds of bees) beside the observation hive (with a glass side lifted) about thirty minutes before dark and let the swarm move in.

Starting with a Frame of Brood

If you plan to start the observation hive with a new mated queen, be sure you have the queen in hand before stocking the hive with brood and bees. At one time or another, you should let the bees make their own queen so you can observe that interesting phenomenon as it takes place.

Equipment needed:
- [ ] Hive tool
- [ ] Smoker or hive bomb
- [ ] Frame grip
- [ ] Observation hive

Step-by-step directions:
- [ ] Carry the observation hive to the hive from which the frame of brood and bees will be removed. Remove the top and one glass side from the observation hive. Put them in a safe place nearby.
- ■ **Dress properly.**
- ■ **Light the smoker** or have hive bomb handy.
- ■ **Enter the hive.** Set the honey supers (if any) on the inverted hive top.
- ■ **Inspect the hive.** As you inspect, look for the frame with the most covered brood. If you are going to let the bees raise their own

queen, you will need a small area of eggs and young larvae from which the bees can make the new queen. Remember that the covered brood are the earliest new bees for the new unit.

☐ Place the frame of brood and attached bees in the observation hive, either through the top or slipped into a slot for the frame end bars from the side, depending on the construction of the observation hive. Replace a frame of foundation or comb in the brood chamber for the frame of brood removed.

☐ Now you will add bees to get additional nurse bees. Tilt the observation hive on its side at an angle of 45° (lean it against something).

One at a time, remove two frames of brood and bees, check carefully that the queen is not on the frame, and shake the bees onto the frame of brood and bees already in the observation hive. Replace the frames of brood, shaken free of bees, in the brood chamber.

Replace any frames of comb or foundation in the observation hive required to fill the frame spaces. If you are going to add a new mated queen to the observation hive, skip the next step.

☐ Replace the glass or plastic side carefully, attach the top with screws or nails, close the exit with a rag, and brush off any bees on the outside of the observation hive. Skip the next step.

☐ If you have ordered and have in hand the new mated queen that you are going to place in the observation hive, proceed with introducing the new mated queen as follows.

Remove the cork from the candy end of the queen cage. Perforate the candy plug with a toothpick, and place the queen cage on the top frame as far to one side as possible so the bees can feed and touch the queen through the screen while they are eating away the candy plug. Fasten the top with screws, close the exit, and brush off any bees on the outside. *Note:* In four days you will move the hive outside to make sure the queen has been released and to remove the queen cage, then you will return the observation hive to where it hung or stood. Go to the next step.

☐ Place the observation hive where it will hang or stand. Quickly remove the exit plug and insert the travel tube, if one is needed to get the bees to the outside. You can wait until after dark to insert the travel tube, which will keep the bees inside until morning. In the morning, all those bees that had oriented to the hive from which the brood and bees were taken will return to the hive, unless it is more than a mile from where the observation hive will be kept.

Starting with a Small Swarm

☐ Capture a small swarm of bees. A swarm about the size of a football or soccer ball is just about right for the 1½- or 2-frame hive. An observation hive started with a swarm can be started on foundation. Store the swarm in a cool place until about half an hour before dark.

☐ About thirty minutes before dark, place the observation hive on the ground near and under the exit where the bees will be leaving the travel tube to the outside. Remove the top and raise the glass on one side of the hive about four to six inches and prop it up. Dump the boxed swarm as close to the observation hive as possible. You may want to set the hive on cardboard or plywood if the area is not smooth ground or concrete. The bees should begin to crawl into the observation hive and onto the frames of foundation or comb. Let the swarm move into the hive over the next thirty minutes. (See Figure 33-1.)

☐ At about dark, most of the swarm will be inside the observation hive. Gently lower the glass, place the top on the hive, and fasten it closed with screws. Close the opening and brush off any bees on the outside of the hive. Carry the hive to where it will hang or stand. Quickly remove the plug from the opening, and insert the travel tube. The bees brushed off the hive before you bring it inside will join the hive when the bees leave the hive in the morning and begin to scent and orient.

REMOVING THE OBSERVATION HIVE FROM THE HOUSE

Occasionally it is necessary to remove the observation hive from where it hangs or stands and take it outside to remove a queen cage, clean up the hive, or shake bees off the frames to keep the observation hive from swarming.

Figure 33-1, Starting an Observation Hive with a Small Swarm. A swarm was dumped beside the observation hive. The glass on one side is being held up by a frame end so the bees can get into the hive. The swarm was dumped about one-half hour before dark. Most of the bees have crawled into the observation hive onto the combs on the inside.

Equipment needed:
- ☐ Several bits of rag
- ☐ Hive tool
- ☐ Smoker or hive bomb
- ☐ Frame grip
- ☐ Queen catcher or small jar
- ☐ Screwdriver

Step-by-step directions:
- ☐ On the night before you are going to remove the hive from where it hangs or stands, get three pieces of rag, each about six inches square. Go to the observation hive, gently remove the travel tube, and close the exit of the observation hive securely with one of the rags. Close the exit of the travel tube with a second piece of rag. Now move outside and close the exit of the travel tube leading to the outside.
- ☐ In the morning when you are ready, ■ **Dress properly.**
- ■ **Light the smoker** or have the hive bomb handy.
- ☐ Remove the observation hive from where it hangs or stands, and carry it outside to the area where the bees exit from the travel tube. Place the observation hive on a piece of cardboard or plywood to open the hive, remove the frames, etc.
- ☐ Remove the top and puff smoke into the observation hive. Now you can remove the queen cage and replace the top, brush the bees from the outside of the hive, and return it to the house. Bees that fly out will go to the closed travel tube.
- ☐ If you are going to clean up the bottom or remove the queen to use her for re-queening, proceed as follows.

  After removing the top and smoking the hive, carefully remove one of the glass or plastic sides of the hive. Puff smoke over the frames. It may be necessary to raise or remove the frames to clean the hive or to capture the queen with a queen catcher or small jar. Use a hive tool and frame grip as necessary to get the job done.
- ☐ Replace the frames in the observation hive, if removed. Replace or lower the glass or plastic side. Replace and fasten down the top. Brush off any bees on the outside of the hive, and return it to where it hung or stood.
- ☐ To reconnect the travel tube:
    1. Remove the plug from the end of the travel tube.
    2. Remove the plug from the observation hive exit, and quickly slip the end of the travel tube into the observation hive exit.
    3. Remove the plug from the outside exit of the travel tube.

KEEPING THE OBSERVATION HIVE FROM SWARMING

Observation hives may swarm during the honey flow. The best way to keep the observation hive from swarming is to remove the hive, carry it out to the big hive, capture the observation hive queen, and shake the bees off the frames. Replace the queen in the observation hive, and return it to where it hung or stood. Reconnect the travel tube to let all the older bees back into the hive. All the young bees will have joined up with the big hive, so the population will be reduced and swarming will be avoided. You may have to repeat the procedure one or two times during the honey flow.

Equipment needed:
- [ ] Several bits of rag
- [ ] Hive tool
- [ ] Smoker or hive bomb
- [ ] Frame grip
- [ ] Screwdriver
- [ ] Queen catcher or small jar

Step-by-step directions:
- [ ] When the hive is overflowing with bees or if you see queen cells being generated, it is time to dump the bees in front of the big hive so the observation hive will not swarm.
- [ ] On the night before you are going to dump the bees, close the observation hive exit and the ends of the travel tube, as directed earlier.
- [ ] When ready the next morning, ■ **Dress properly,** then remove the observation hive from where it hangs or stands, and place it beside the big hive.
- ■ **Light the smoker** or have the hive bomb handy.
- [ ] Place a ramp up to the entrance of the big hive.
- [ ] Remove the top of the observation hive and puff in some smoke. Gently raise the glass or plastic on the side where the queen is seen. Puff in some smoke. Remove the glass and the frame, if necessary, to capture the queen in a queen catcher or small jar. Place the queen in a shady spot.
- [ ] Remove the glass from one side, and then remove the frames and shake them on the ramp in front of the big hive. Shake or brush the bees from inside the observation hive, and replace the frames. If you find any queen cells being generated, tear them off or otherwise destroy them at this time. Replace the glass and release the queen back into the observation hive. Replace and secure the top. Brush

off any bees on the outside of the observation hive, and return it to where it hung or stood.
- ☐ Reconnect the travel tube as directed earlier. You may find that you have to move bees to reach the plug in the outside exit of the travel tube. After you shake the bees in front of the big hive, all the old bees will return to the outside entrance of the observation hive and pile up there, waiting to get back in. Next time, fasten a safety pin (with a string attached) to the rag when you plug the outside exit of the travel tube, and you won't have to dig among the bees to find and remove the plug.

After you shake the bees in front of the big hive, all the young bees will run into the big hive and become a part of that hive. In this way you will have reduced the crowding in the observation hive and avoided swarming until the hive becomes crowded again. If it does, perform this manipulation again.

If you forget to plug the outside exit of the travel tube, you may return to find the bees jammed into the tube so tightly that they cannot move when you reconnect the tube to the observation hive. If this occurs, you must remove the bee plug or they will all perish. Here is the trick to open a plugged travel tube:
1. Remove the plug from the travel tube where it enters the observation hive.
2. Take a deep breath, place your mouth on the end of the travel tube, and blow hard. The plug of bees will go shooting backwards out of the travel tube, probably shouting *whee!* or exuding a pheromone, which implies the same thing.

WHAT YOU CAN SEE IN THE OBSERVATION HIVE

- Bees feeding larvae, cleaning and polishing cells, storing honey, and resting.

  During these activities a bee crawls into a cell so you see the rear of the abdomen protruding from the cell. Probably a difference in the time spent in the cell or in the depth of movement into the cell best indicates the activity being performed. Feeding larvae and storing honey takes less time than cleaning and polishing a newly vacated brood cell. Bees apparently do rest in empty cells; when at rest, they are deep in the cell, with just the tip of the abdomen showing.
- Bees feeding other bees and field bees transferring nectar to a "house" bee.

  These activities occur when two bees are facing each other, one with its tongue extended into the mouth of the other. Bees do a lot

of food tranferring from bee to bee within the hive. There is probably a signal which initiates the transfer.
- Bees capping honey cells and brood cells.

    This activity is more easily seen in cells containing honey. Nectar becomes honey when its water content is reduced to about 17 percent moisture, and the bees know when this occurs. As soon as the honey is ripe (17 to 24 percent moisture), the cell is capped. Look for cells that have been partially capped. When you find one that has a bee working around the edges with its mouth parts, you are seeing a cell of honey being capped. The larval cappings are not the same as those on the honey. The larval cappings must be permeable to circulate air and vent moisture. If you see a bee working on a partially capped brood cell, you are seeing a brood cell being capped.
- Bees emerging from brood cells.

    Look in the brood area for a covered cell that has a small opening in the cap. This may be a cell from which a bee will be emerging or one that is being capped at the end of the larval stage. If you watch the hole for a bit, you may see an antenna sticking out. If you see this, the bee inside is cutting away the capping and will be emerging within a few minutes.

    Bees do not back into cells. If you see a bee with its head sticking out of a cell, watch and you will soon see the bee pop out of the cell like a cork from a bottle. The hair of the new bee will appear to be damp and its coloring will be different from the older bees nearby.
- Bees secreting wax and making comb.

    You will not see bees actually secreting wax flakes, and you will not see much comb being made. However, you will see the end result: frames of comb—all but the foundation sheets—that are made from the wax flakes secreted by the wax glands of thousands of bees. The beekeeper provides the frames with foundation that has been stamped with cell bases so the bees will make the cells all the same size—just the size in which the queen will lay a fertilized egg to produce a worker bee. Drones (male bees) are reared in larger cells.
- Bees doing the communications dance.

    This is one of the easiest activities to spot. When you see a bee moving on the frame, wagging her abdomen vigorously from side to side, you have spotted the wag tail dance, which is part of the communications or recruitment dance.

    Watch the wag tail run; see the bee turn either right or left, return to where the run was begun, and make another wag tail run. At the end of this run, the bee turns in the opposite direction, returns, and repeats the wag tail runs, making the figure-eight pattern several times. Usually she then moves to a new area and dances again. The figure-eight dance signifies that nectar or pollen is available between roughly 150 feet and several miles from the hive.

The wag tail dance specifies both direction and distance. To indicate direction, a wag tail run toward the top of the frame says "go from the hive into the sun." The dance toward the bottom of the frame says "go from the hive away from the sun." A dance at 60° counterclockwise from the top of the frame says "leave the hive and fly 60° to the left of the sun," etc. The wag tail dance indicates distance by the number of wag tail runs in a certain time span and perhaps by the length of the run. Short runs rapidly repeated—say, ten runs in fifteen seconds—means that the food is in the direction indicated at about 300 feet. Two longer runs in about fifteen seconds indicates that food is in the direction indicated at about 15,000 feet (three miles or 5000 meters).

Bees do a round dance for nectar or pollen that is closer than 150 feet from the hive. In the round dance, the bee makes almost a 360° turning run (without wagging), reverses direction, makes a 360° turning run in the opposite direction, and continues to repeat the circle with direction change. The circle or round dance says "search within 150 feet, and find the scent you smell on me," but does not tell the direction or distance more accurately.

While the bee is dancing, other field bees are smelling and sometimes being offered a taste of the nectar. Thus, when they leave the hive, they know the direction, distance, and scent of the pollen or nectar for which they search. Bees do these dances to indicate the location of nectar or pollen, both of which are vital to the well-being of the hive.

- Bees storing and packing pollen.

After the bees carrying pollen have completed their dances here and there on the frames, they go in search of a cell in which to store the pollen or one in which pollen has been stored earlier. After looking in several cells and being satisfied that one is good, the bee steps ahead, sticks its hind legs into the cell, and begins to rub them together, knocking the pollen balls off into the cell. The bee then moves out and away. If you are sharp and look quickly, you may see the pollen balls on the bottom of the cell. Very quickly, however, a house bee will stick its head and thorax into the cell and begin to pack the pollen firmly down into the cell. Nectar is mixed with the pollen to help it pack nicely and perhaps to help preserve it. A certain amount of fermentation takes place as the pollen ages.

The bee who finished dancing but was carrying nectar will go off and transfer its load of nectar to a house bee. The dancing bees are now ready to return to the field for additional loads.

- Bees cleaning the hive of debris, dead bees, wax moth larvae, etc.

This cleaning activity probably can be seen best at the entrance to the hive. Watching the bees coming out of the hive, often you can see one carrying a bit of wax or wrestling out a dead or deformed bee. Occasionally you will see a wax moth larva being carried out. The bees try to carry away all unwanted material and drop it some distance from the hive. On the frames, you might see a bee with a bit of wax or one carrying or dragging a dead bee.

- Bees forming a circle around the queen, each facing the queen.

This is one of the things you look for when searching for the queen. It is a characteristic pattern when the queen is present in the area, particularly if she is resting after a period of egg laying. You will recognize the very definite pattern when you see it—the queen standing with many bees facing her, forming a ring in which she is the center. The queen is harder to locate when she has her abdomen in a cell while laying an egg.

- Bees feeding, caressing, and preening the queen.

Watch the bees in the circle facing the queen. Many will be touching her with their antennae, some may be licking her, another may be feeding her, and others may be working on her body with their mouth parts as though scratching her back. As the queen moves along, other bees will turn toward her as she passes. If she pauses for a while, the ring of bees will again form around the queen.

- A queen laying eggs.

This is always fascinating to watch. The queen moves along, looking into the cells as she goes, until she finds a cell that is empty and polished. Some queens stick their heads right down into the cell. If the cell suits her, she steps ahead a bit, gets a good grip with her feet, bends her abdomen, and lowers it into the cell. She may adjust her feet or turn a bit in order to better set the egg in the center of the cell base. After about fifteen seconds, she moves forward out of the cell. If many cells are available, empty, and clean, as she steps forward from laying in one cell, the queen will look into another cell, step ahead, get a firm grip, bend her abdomen, and lay another egg. This may go on in one cell after another until the queen decides to rest and take food.

- The contrast between drone brood cells and worker brood cells.

In the new hive, the drone brood usually first shows up along the top, bottom, and sides of the brood comb frames. It is here that you will see transition cells, which are made smaller or larger than the other cells in order to join the perimeters to the top bar, bottom bar, and frame side bars. Unfertilized eggs are laid by the queen in the larger cells, because the drone larvae are larger than the worker larvae.

In addition, the bees make more room in the cell by extending the brood capping of the drone cell. Seen individually, a capped drone brood cell has a characteristic, protruding cover that looks like the end of a .22 caliber bullet. The covering protrudes ⅛ to ¼ inch above the normal, covered worker brood cells. (See Figure 26-1.)

Older brood comb may have areas of comb that have been modified and have large areas of drone cells. Drones are necessary for the morale of the hive, in case something happens and a new queen is required. If there are no drones in the area, there is no way for a virgin queen to mate, and the hive will die out.

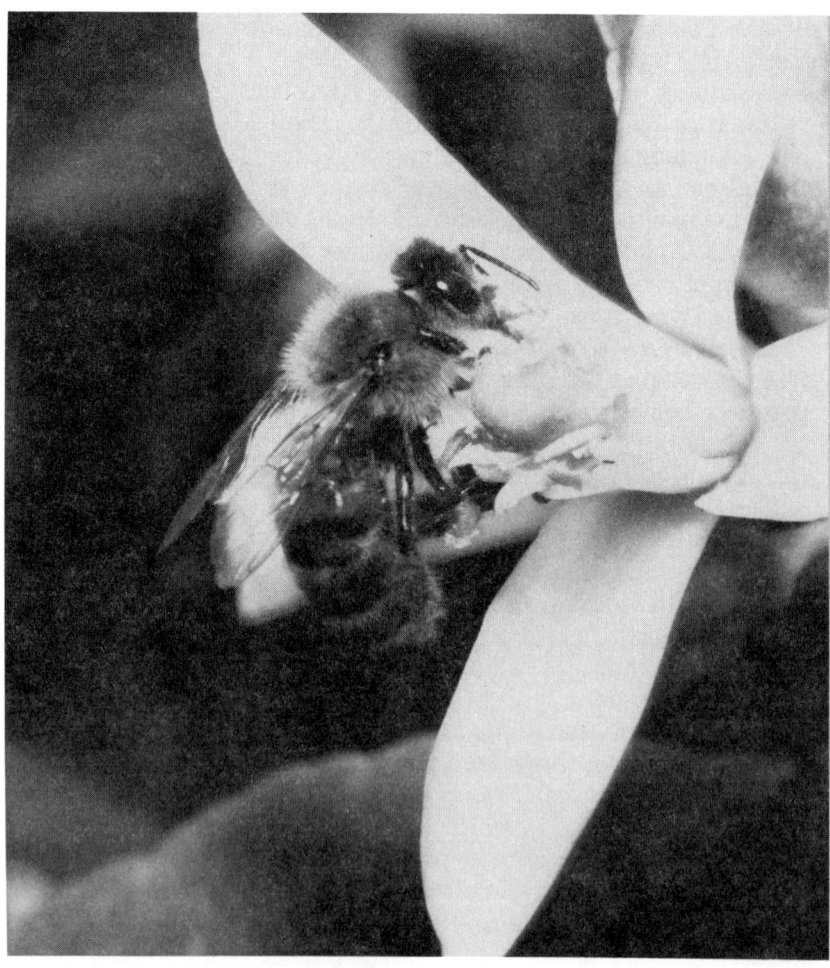

UNIT 34

Pollen Traps

Pollen is the bees' natural protein food. Honey is the carbohydrate. Younger bees, who produce royal jelly, must ingest large quantities of pollen, which is somehow converted or metabolized in the bee's body to produce large quantities of royal jelly. Royal jelly is fed to the very young larvae, developing queen larvae, and the laying queen for the heavy production of eggs during the spring build-up and during the honey flow.

Pollen traps are used to gather pollen from the legs of the field bees. Many types of pollen traps are available and many more are being invented or redesigned, because the production of pollen as a cash crop has become more important in the last few years. Two types of pollen traps are discussed here. First is the pollen trap that hangs on the front of the hive and can be moved from hive to hive. Second is the pollen trap upon which the hive rests, without the bottom board. The latter model becomes a part of the hive.

You can collect pollen during the year to feed back to the bees in the early spring in those areas where there is a scarcity of early pollen. You can also collect pollen to use as a high protein supplement in your own diet. In some areas with larger populations, the pollen you collect can be sold to health food stores as a protein supplement or for use by those suffering from pollen allergies (hay fever).

HOW A POLLEN TRAP WORKS

Regardless of the model or placement of the pollen trap, the field bees must climb through a single or double six-by-six mesh screen or perforated metal sheet in order to get into and out of the brood chamber. As the bees crawl through the mesh, the bees with larger pollen pellets on their legs lose the pellets, which fall through an eight-by-eight mesh screen into a drawer or tray. The drawer can be removed or detached to recover the pollen pellets for storage. The bees are not dumb, and if the pollen trap is left on continuously, they begin to bring in smaller loads of pollen, which are not removed as they crawl through the mesh screen. Normally, the beekeeper will take one day's pollen, wait three days, and take the next day's pollen, continuing to take pollen every fourth day during the season.

TWO TYPES OF POLLEN TRAPS

1. Pollen traps that hang on the front of the hive.

    This type of trap hangs or is supported on the hive between the upper and lower brood chambers. One model hangs from a rim, which raises the upper brood about ¾ inch, leaving an opening at the front of the hive between the two brood chambers. The rim extends far enough beyond the hive so the pollen trap, when in position, hangs over the rim-made entrance, and the bees must pass through the mesh of the trap to get into the hive. The pollen is removed from the bees' legs and drops into a drawer for recovery at the end of the day. Using this model, you must close the hive entrance completely so the bees can come and go only through the rim-made entrance between the two brood chambers. (See Figure 34-1.)

    Another model uses a setback between the lower and the upper brood chamber. The normal hive entrance is closed, and the bees must use the offset entrance. An angled piece of tin closes the offset entrance at the back of the hive when the two supers are offset. The pollen trap has a 1/16-by-¾-inch plate on each side that slips between the brood chambers. When the trap is in position, it mates with the offset brood chambers in such a way that the bees must go through the pollen trap mesh screen and lose the pollen pellets, which fall into a drawer.

    Both of these models can be moved from hive to hive once rims or angled tins are installed in several hives.

2. Pollen traps that become part of the hive.

    This type of pollen trap is more expensive, and a separate pollen trap is required for each hive that is used to trap pollen. The pollen trap becomes the new bottom board of the hive. The hive rests on the pollen trap. Some models have a pollen drawer that can be removed from the back of the hive, and others have a drawer that can be removed from the front of the hive. Most pollen traps of this sort have a removable six-by-six mesh screen arrangement that can be reversed, or a movable, perforated metal sheet. In one position, the bees are forced to crawl through the mesh screen, which removes

Figure 34-1, Hanging Pollen Trap. Left: The hive entrance has been closed and the rim for holding the pollen trap inserted between the two brood chambers. The bees fly from the opening between the two brood chambers when the trap is not in place. The pollen trap (sitting on top of the hive), when in place, is held by the rim extending in front of the hive, and the bees must go through the trap to get into the hive. Below: A view into the pollen trap, showing the mesh that strips away the pollen pellets as the bees crawl through. The pellets fall through the finer mesh screen into the drawer of the trap.

the pollen pellets. In the reverse position, the bees enter above the mesh and go directly into the brood chamber. (See Figure 34-2.)

For the beekeeper with a few hives, the pollen trap that hangs between the brood chambers of the hive is probably the best investment. One pollen trap can be used on up to four hives if you duplicate three additional rims or three angle plates. Using this system, you

Figure 34-2, A Pollen Trap That Becomes a Part of the Hive. Above left: The pollen trap as it looks when it is set under the brood chambers of the hive. The pollen trap can become the bottom of the hive, but in this photo, the bottom board has been closed off and the pollen trap set on the bottom board. The bees enter the hive through the entrance in the pollen trap. The plastic tubes allow the bees to remove most of the dead and also allow the drones to leave the hive. Above right: Rear view, showing the perforated screen (that removes the pollen pellets) that the bees must crawl through to get into the hive. Notice the holes in the end of the hive leading to the plastic tubes seen in the front view. Below: Rear view, showing the perforated screen being held up to allow the bees to enter the hive without losing pollen. The screen seen in this shot is the finer mesh screen through which the pollen pellets fall (when pollen is being collected) into the pollen drawer, which, in this model, is removed from the rear of the hive.

can make the pollen trap work one day on each hive, and on the fifth day you can start over again. You can take pollen from four hives all during the season and probably gather fifteen to fifty pounds of pollen.

Pollen should be collected each evening and the pollen traps moved to the next hive if you have the hanging pollen trap. Pollen should be collected before it rains, because most pollen traps let water drain into the pollen drawer, and this will spoil the pollen. Pollen should be stored in plastic bags and placed in the freezer until needed or processed for sale.

POLLEN FOR HUMAN CONSUMPTION

Pollen for human consumption must be cleaned and dried. The pollen must be toasted at about 150° F for an hour or so until all the moisture has been removed, and then it can be packed in glass or plastic jars so it will not mold or spoil. You must clean the pollen by picking out the legs, wings, and other things that are collected with the pollen. Most pollen traps gather bee parts because the bees try to carry out dead bees while the six-by-six mesh screen is in place; they tear off wings and legs trying to pull the dead bees through the mesh screen. Most Americans are not too keen on getting the fiber and protein of bee parts along with the pollen protein, so the parts must be removed before you sell the pollen to customers or health food stores. Part of the high cost of pollen as a food source is the cleaning. The best way to clean small lots of pollen is with your fingers or a pair of tweezers.

POLLEN FOR FEEDING THE BEES

Pollen used for feeding back to the bees can be kept in the freezer until needed in the spring. If you have enough pollen to feed the bees natural pollen patties, do so. If you do not have enough, mix the natural pollen with pollen supplement. The more natural pollen in the pollen supplement, the more readily the bees will take it.

UNIT 35

The Solar Wax Melter

Solar wax melters are rather expensive when purchased. However, with plenty of sunshine, even a jury-rigged solar melter can reduce cappings and wax scraps into blocks of salable wax or wax blocks that can be refined further for use in making candles for sale or as gifts. (See Figure 35-1.)

Write or visit a university extension service or an agricultural college in your area, and try to obtain plans for making a solar melter, an observation hive, and a pollen trap. If you are handy, you can save a fair amount of money by making your own accessories, and you may be able to sell some of them through your local beekeepers' association. Perhaps you could contact the local high school woodworking instructor, and one or more of these accessories could become projects for students who could do the work if the materials and plans were provided.

THINGS TO CONSIDER WHEN BUYING OR BUILDING A SOLAR MELTER

1. Glass is much better than any plastic for the top of the solar melter. Thermopane or a double thickness of glass is best, but when the sun is shining in the warm season, a single thickness of glass will melt wax of any sort very quickly, and the metal pans for gathering the wax can become almost too hot to handle.
2. Consider a top that can be opened from the front or side as well as removed completely. Given a choice, having a top that can be removed is preferable to a hinged top.
3. The inside dimensions should be wide enough and long enough to accept two deep frames side by side. The depth of the melter, from the glass top to the metal melter tray, should be deep enough to take a good thickness of cappings and wax scraps. The melter should

Figure 35-1, A Solar Wax Melter. A new wax melter. The top is covered with a plastic material. Wax is caught in small bread pans at the base of the metal pan on which the wax scraps are placed. Two blocks of wax were placed on the corner of the melter to show the variation in wax colors. The lighter is from honey cappings, and the darker is from wax scraps and brood comb.

be wide enough and long enough to hold queen excluders for melting the burr comb from them when it becomes necessary. Do not place plastic queen excluders or plastic frames in the solar melter, for the temperature inside gets hot enough to deform most plastics.

4. The solar melter should be deep enough that a large container for gathering the melted wax can fit under the metal melter tray. It need not be overly deep, but it should be able to take a pan that has a large volume. There should be a door on the end of the solar melter to allow the removal and replacement of the wax-gathering receptacle. This door makes it possible for the solar melter to be shorter, because the wax receptacle can slide under the melter tray rather than having to slip down in front of it.
5. The solar melter should be mobile so that it can be rotated and moved to get sun early in the morning and late in the afternoon. The author has modified an old lawn mower body to hold the solar melter so it can be moved around the yard.
6. When melting frames of brood comb or wild brood combs, place an old queen excluder or a piece of ¼-inch hardware cloth in the melter to hold the cocoons off the metal melter tray. This way the cocoons will not absorb too much of the wax, and the wax can flow out from under them.

MELTING CAPPINGS AND SCRAP WAX

Fill the area between the metal melter tray and the glass at the top as full of cappings and scraps as possible. Remember that the cappings will contain about three pounds of honey to one pound of wax. As the cappings warm in the melter, the honey will begin to run into the wax container. As the wax melts, it will float on top of the honey. When cool, the hardened wax can be lifted off the honey, and the honey can be used for cooking or feeding back to the bees, if your hives are disease-free. Any small or thin pieces of wax removed when honey was taken during the melting can be placed in the melter later and melted into larger blocks. Wild comb containing honey can be melted down to recover the honey and wax.

If you extract late or during the winter and lose the sun power, save your cappings in buckets until the sun is available to melt the cappings in the spring. Cut the cappings and granulated honey from the bucket, and place them in the melter. The granulated honey will melt, and the wax will form a layer on top of the liquid honey.

As the wax melts, a certain amount of residue will float down into the pan, such as pollen, bits of brood cell cocoons, bee parts, etc. Wax is lighter than most of the debris and floats above it. When the wax solidifies, much of the debris is caught in the lower level of the wax block. To clean away this residue on the bottom of the wax block, set the block out in the sun with the residue facing the sun. In about two hours, the wax will soften and the residue can be scraped away with the hive tool. Toss the wax containing the residue back into the melter, and remelt it with the next batch of scraps or cappings.

REFINING WAX FOR CANDLE-MAKING

Perform the step outlined in the previous paragraph to remove the residue from the bottom of the block of wax. The next step is to place the wax block directly in two or three quarts of water, and heat it on the stove until it is completely molten to remove any traces of honey that may have been trapped in the wax block. The water will dilute or absorb any honey in the wax. Let the wax cool into a solid block. If traces of residue show on the bottom of the block of wax, set the block in the sun for several hours, and scrape away the residue. Place the waxy residue in the melter.

Your wax should now be ready to be made into candles. For reaching wax temperatures under 200° F, melt the wax in a double boiler. For reaching wax temperatures above 200° F, melt the wax in a can or pan, starting with a low heat, until enough liquid wax is available in the container so as not to scorch the wax. Use a wax or candy thermometer in all wax meltings. If using plastic molds, do not heat the liquid above 175° F. If using metal candle molds, heat the wax to about 210 to 220° F. If you are sand casting, heat the wax to about 240 to 260° F.

EXCHANGING WAX BLOCKS FOR FOUNDATION

If you have at least twenty-five pounds of wax blocks and can get them to a bee supply house, you can exchange your wax, pay a make-up charge of so much a pound, and return home with twenty-five pounds of foundation. Most bee supply shops will buy wax for somewhat less than the bee supply houses pay and later exhange it for foundation when they have several hundred pounds to ship away.

UNIT 36

Looking Back and Looking Ahead

LOOKING WAY BACK

About 100 million years ago, when the first flowers began to appear on plants and trees, the ancestor of the honeybee began to evolve, using the nectar and pollen of those flora for subsistence and keeping the species from extinction.

For the next eighty to ninety million years, the bees had no worries other than fire; drought; and various insects, birds, and mammals. Year after year, the primary objective of the honeybee was the same as that of every species at the time; namely, to exist and generate new units to keep the species from becoming extinct. The honeybees did this by building up the population each spring, swarming, producing honey for winter stores, and clustering through the colder days and nights of winter. Year after year this continued with the honeybee adapting as necessary to its hereditary evolution and the changing environment. It appears that the bee has changed little in appearance over the last few million years.

Somewhere in those last ten to twenty million years, an obscure, unattractive mammal began to develop a brain. Humans started their dominance over the world and everything in and on it. They began to rob the bee trees for the sweets and grubs and pollen. It is probably due to the stinger of the bee that humans have not caused the bee's extinction.

LOOKING SOME DISTANCE BACK

About 10,000 years ago, when humans began to leave behind some written records on cave walls, clay tablets, etc., we began to learn about bees and honey. It is known that as early as 5000 B.C., honeybees were being kept in skep-type baskets and clay pots. Honeybees at this time were indigenous only to Europe, Africa, and Asia. Migratory beekeeping came early to Egypt in the form of floating apiaries on the Nile River. The ancient Greeks and Romans left behind records of bees and beekeeping.

LOOKING BACK IN THE AMERICAS AND AUSTRALIA

Honeybees—the mean, black bees from northern Europe—were brought to the American colonies soon after colonization began, and were probably here by the mid 1600's. It is assumed that the honeybee was brought into South America sometime in the 1600's or early 1700's as colonization progressed in that area. Honeybees were brought into Australia around 1822 and New Zealand around 1842.

Honeybees were called "the white man's flies" by the Indians in the American colonies. It is interesting to note that swarms from the original hives brought into the colonies had reached across the forested areas to the plains by the time the pioneers moved west. Some pioneers made money by cutting down bee trees and selling the honey. Apparently the Great Plains and the Rocky Mountains stopped the bees from going to the West Coast, because honeybees were first brought to California and the West Coast in 1851, being shipped around South America. After being loaded off the ship, those hives were transported to San Jose, California, and a plaque has recently been erected to memorialize that first apiary site in the state.

Since 1851, when the movable frame hive as we know it today was invented, commercial beekeeping has been possible. Following the movable frame hive there appeared in rather rapid succession the extractor, the smoker, wax foundation, the bee escape, and most of the other beekeeping items and accessories now available. Thousands of different accessories and hives have been tried over the intervening years, some of which are with us now; the rest have just not worked out.

The honeybee that made commercial beekeeping possible was the Italian bee, which was brought into the country in the mid 1800's. Some years later, queen rearing and the shipment of Italian queens began; beekeepers were "Italianizing their hives," which is to say, re-queening to get rid of those mean black bees.

LOOKING AHEAD

Looking into the future, we may find that the Africanized bee moves into the lower half of the United States and that the Varroa mite also may come into the country, because it is now in many areas of the world. Of the two, the Varroa mite probably will cause more problems. The greatest problem is how to kill an insect living on another insect without killing both. Perhaps by the time the Varroa mite arrives, some prevention or control will have been developed. It is possible that we will be able to feed to the bees a medicated solution that can poison their body fluids sufficiently (without killing the bees) that the mites sucking those fluids are killed.

The Africanized bee will probably have to be controlled on a hive-by-hive basis by removing the Africanized queen and re-queening with one of the gentle strains now available in this country. The queen-rearing industry may have to move into the northern areas of the country for some years if the Africanized bees invade the queen-producing areas in the South and in California.

In the more distant future, one can expect to see more breeding work done with bees to make them more gentle or more productive. It is possible that with biological engineering, special bees can be produced to do the jobs they cannot or will not do now. It may be possible to produce a bee without a stinger. Such tampering might cause the extinction of the bee, when anyone or anything could attack the hive and the bees could not defend their home. It is possible that almost everyone would keep bees if the bees had no stingers, except those who prefer to take what someone else has worked for, particularly if the bees could not protect themselves.

THE HERE AND NOW

Even with bees having stingers, beekeeping can be the hobby for almost anyone with a bit of backyard and the interest or desire. Read, observe, learn, and think about the behavior characteristics of the bees. Those behavior characteristics are the mental tools that permit the more observant beekeeper to manage the bees more easily at the proper time and to make the management of the bees possible with the least amount of work, fear, or pain.

Over the years as you continue to keep bees, you will learn more and more about the bees by making mistakes, by accident, and by experimenting.

Consider buying or making an observation hive within the next few years, and let your family and friends learn the fascination of beekeeping or bee viewing in the safety of your own home. In the North, you will have to restock the observation hive each spring with a frame of brood and let the bees make their own queen before your very eyes, or buy a mated queen to introduce into the frame of brood in the observation hive. With the observation hive you will always have a laying queen to introduce into one of your big hives in case you need a queen. In the South, you can often keep the observation hive going right through the winter by feeding the bees. There is quite a lot of heat loss through the glass, and the hive population tends to deplete considerably before spring.

Consider teaching a class in beekeeping after you have a few years' experience. Talk to school groups, at library gatherings, and garden clubs. There is no better way to learn more about bees than to teach or talk about them. You will soon find that it is hard to cover as much as you would like in an hour session. If you ever lose your notes, just open the discussion by asking for a question; the questions will come, one after the other, and your talk can be almost extemporaneous. One great teaching or talking aid is your observation hive. Close it up and take it along when you go to talk to groups. Word will get around, and your talks will be in demand. The author keeps several observation hives going in the backyard during the spring, which are rented to the schools for a two-day period. The observation hives are closed off, sent to schools, observed by the teachers and students, and are often viewed in many classrooms before being returned to the backyard. After being turned loose to fly for several days, the bees are ready to visit another school.

If you can find at an auction, flea market, or second-hand bookstore one of the early books on beekeeping such as *The ABC and XYZ of Bee Culture* or *The Hive and the Honey Bee* published around 1913, you will be in for some interesting reading. You will also find that the beekeepers of that day actually were learning in the apiary most of the things we know today. Almost everything imaginable in hives and management were tried over the years, and many of the things tried did not work or were too much work. Many misconceptions abounded then, and many of the misconceptions are still with us. Years ago, the bee magazines were published twice monthly and were full of more questions than answers.

Over the years, many universities and the Department of Agriculture Bee Research Stations have been carrying out experiments with bees and year by year are bringing to light the behavioral characteristics of the bee. The bees are a puzzling conundrum wrapped in an enigma, which makes Rubik's Cube seem like child's play.

Read as many books on beekeeping as you can. Trade off with beekeepers in your area. See if your beekeepers' association will start a lending library. Buy for your own library those books that you find most interesting. You should select one or several bee magazines and subscribe. Through bee magazines you will stay current with the latest developments in apiculture. The bee magazines are about the only source available for package bee and queen bee producers. In the advertisements you will see, as subscribers have seen over the years, many things that are supposed to work. Some will fail and fall by the wayside as so many things have, but there may be something that will really benefit beekeepers, as this author hopes his books will.

GLOSSARY

American Foulbrood (AFB) An infectious brood disease of honeybees caused by the spore-forming germ *Baccillus larvae*, controlled by the medicine, Terramycin®.

Apiarist One who keeps bees.

Apiary A place where beehives are kept.

Banking queens Holding one or many queens in queen cages inside a queenless hive for several days to several weeks while waiting to sell or use the queens for re-queening.

Bee blower A high velocity blower (up to 100 MPH) used to blow the bees from the honey supers. A nozzle is moved along the frame tops blowing the bees from between the frames and the sides of the honey super.

Beehive (or hive) The standard sized boxes (supers) with frames and foundation which house a colony of bees.

Bee supply house As interpreted for this volume, a company that manufactures or constructs most of the bee supplies it sells (some for national distribution), with warehouses in strategic locations.

Bee supply shop As interpreted for this volume, an individually owned retail shop specializing in the sale of beekeeping supplies.

Bottom supering The hive manipulation of removing the present honey supers on the hive and adding the empty honey super or supers on the queen excluder, and then replacing the honey supers removed on top of the empty honey super or supers just added.

Brood area That portion of the hive that is reserved for and in which brood rearing is exclusively carried on.

Brood frames Frames in which the queen lays the eggs and the metamorphic cycle of the bees take place in the comb cells. These frames are found in the brood area.

Burr comb The buildup of extraneous comb between frame top bars or between top bars and bottom bars between two supers. Depending on its location it may also be called brace comb, ladder comb, crosscombing, etc.

Cappings The wax coverings of honey cells after their removal from combs of honey during the extracting process.

Cell (or brood cell) One of the mass of hexagonal cells built by the bees. Occasional reference is made to a brood cell, pollen cell, drone cell, or queen cell (a special cell).

Cluster The gathering together of all the bees in a hive for warmth when the temperature drops below 55° F. A swarm clusters (gathers together) on a branch, post, etc., as a single unit until they find a new home or are captured.

Communications dance A wag tail dance performed by the worker bees to indicate the distance and direction to a nectar or pollen source.

Colony of bees All of the individual bees, of all castes, who make up the single unit or hive, headed by one queen. The term is often used to denote the hive or beehive.

Comb or honeycomb The mass of hexagonal cells built by the honeybees on foundation in the hive to contain the brood, pollen and honey stores.

De-queen To search for, find, and remove the present or old queen in the hive prior to re-queening with a new, young, mated queen.

Drawing comb The process whereby beeswax, secreted by wax glands on the abdomen of the honeybee, is removed by spines on their mid-legs, brought to their mouths where it is manipulated and sculptured into comb.

Drone The male of the honeybee species, who has no stinger and does no work. He is important only for mating with a virgin queen, which is vitally important for survival of the species.

Drone laying queen A queen who has run out of sperm, or a queen who never mated—capable of laying only infertile (drone) eggs.

Enzyme A complex organic substance produced in plants or animals (insects) capable of producing chemical transformations in other substances by catalytic action. In nectar, the complex sugar (sucrose) is changed to simple sugars (dextrose and levulose) primarily by the enzyme invertase, which is produced in the bees' body.

European Foulbrood (EFB) An infectious brood disease caused by the germ *Streptococcus pluton*. The disease is controlled by the medicine, Terramycin®.

Field bees Bees over twenty days of age who spend their daylight hours gathering nectar, pollen, or propolis.

Floral sources All of the plants and trees which bloom in an area during the year.

Foundation Beeswax sheets impressed with cell bases on which the bees will draw comb.

Frame The movable units within the super (usually ten frames per super) into which the sheets of foundation are securely attached.

Granulation of honey The process by which liquid honey becomes tiny or heavy crystals of honey, whether in jars or in the combs. This results in a "sugared" texture and appearance. There are only a few honeys that will not granulate within the first few months after being extracted or stored.

Guard bee One of a number of bees about eighteen to twenty days old whose job is to guard the hive entrance against robber bees and other predators.

Honey The sweet liquid produced from nectar gathered by bees, processed or elaborated in the bee's stomach, and stored and evaporated to contain sixteen to twenty percent moisture.

Honey flow That season when the greatest number of blossoms of all floral sources are competing for pollination services of insects who provide each service.

House bee Those bees, one day to twenty days of age, who perform all the in-house chores of the hive and do not gather nectar or pollen.

Laying workers In the absence of queen substance (no queen in the hive), some worker bees can lay eggs. Since they do not mate, eggs laid by them are infertile, developing only into drone or male bees.

Mating flight The single or multiple flights made by the virgin queen for the purpose of mating, which is accomplished only in flight. Sperm from a number of drones is stored in the queen's body in a receptacle called the *spermatheca*.

Metamorphic cycle of the bee The phases of the metamorphic cycle in the honeybee include egg, larva, pupa, and adult. The elapsed time of the cycle for a queen is sixteen days; for a worker bee, twenty-one days; and for a drone bee, twenty-four days. All phases of the cycle are accomplished within the same cell in which the egg is laid.

Nectar A sweet liquid secreted by the nectaries of a plant. When gathered by the honeybees it is converted into honey.

Nectaries The plant glands that secrete nectar.

Nosema Disease A disease of adult bees caused by the spore forming protazoan *Nosema apis*. The disease is controlled by the antibiotic fumagillan. Medicines available which contain fumagillan, used to control the disease, are FUMADIL-B® and NOSEM-X®.

Observation hive One or more frames of brood, bees, and queen in a case with glass or plastic sides, having access to the outside. Every activity of the standard hive takes place in the observation hive except storage of excess honey.

Odd hive As used in this volume, any hive that is located in a structure or any odd movable unit, such as a bird house, barbeque, trunk, chest, etc.

Orientation flight The daily practice flights of young or house bees from seven to twenty days old. The thirty to forty-five minute flight of the young bees in a hive occurs at about the same time each day, after which the bees return to their in-house duties.

Package of bees Usually two or three pounds of bees placed in a wooden or cardboard package for shipment from the package producer to a beekeeper by mail. Package bees are shipped with a young, mated queen in a cage inside the package.

Paradichlorobenzene (PDB) A white crystalline compound $C_6H_4Cl_2$, used by beekeepers as a fumigant against wax moths.

Pollen Grains produced by the anthers of flowering plants. Pollen brought to the hive is used and stored as food for the larvae and bees.

Pollen patty A round, thin patty made by mixing pollen substitute or natural pollen with a 2-1 sugar solution until it is the consistency of cooky dough.

Pollen substitute A powdered mixture, generally of soyflour, brewers' yeast, and powdered milk that is mixed with a sugar solution and fed as a pollen patty.

Pollen trap A device placed on the hive which forces the bees to crawl through a six-by-six mesh screen or a perforated metal strip which removes the pollen pellets from the legs of the bees and gathers them in a drawer for use by the beekeeper.

Pollination To place pollen on the stigma of a flowering plant.

Propolis A brownish resinous material collected by honeybees from buds of trees and used as cement or glue within the hive to seal cracks and small openings; often called bee glue.

Queen or honeybee queen The only fully developed or complete female in a colony or hive of bees. When mated, she lays fertile eggs to produce worker or queen bees, or infertile eggs to produce drones.

Queen cage A small cardboard box; a small perforated plastic container; or a small hollowed wooden block covered by screen, used to hold a queen bee and several worker bees when sent through the mail or placed in a package of bees for shipping. The queen cage is also used for introducing the queen into a queenless hive.

Queen cells A special cell made by the bees for the rearing of a queen or queens. The queen cell hangs vertically from the face or near the bottom of the brood comb.

Queenright hive A hive that has a laying queen.

Queen succession The replacement of an old or failing queen by the bees who raise a new queen. After the new queen mates and begins to lay, the old queen disappears.

Re-queen To replace an old or failing queen. The beekeeper de-queens the hive and introduces a new, mated queen.

Robbing The stealing of honey from one hive by bees from another hive.

Royal jelly The highly nutritious secretion of the hypopharyngeal glands of the honeybee. Fed to all young larvae for two days, to the queen larvae during the entire larval phase, and to the queen during periods when she is laying heavily.

Scout bees A certain number of bees who leave the hive early each morning to search for nectar and pollen. Most of the field bees in the hive do not go out until the scouts return and perform the communications dance.

Starter strip A one to three inch deep strip of foundation attached at the frame top from which the bees hang and *start* drawing natural comb.

Swarming A natural phenomenon by which the bees increase the species. The old queen and about half the bees leave the hive to find a new home. Now two hives or colonies exist to increase the species.

Supers The standard size boxes used by beekeepers for brood rearing and honey storage.

Top supering The hive manipulation of adding an empty honey super or supers on top of the honey supers presently on the hive.

Uniting hives Combining two different hives to form a single hive. To prevent fighting between bees of the two hives, the beekeeper places a sheet of newspaper between the two hives.

Uniting swarms Combining a swarm with an existing hive or another swarm in a single hive. A single sheet of newspaper keeps the bees from fighting among themselves.

Virgin queen A young queen who has not made her mating flight.

Wax glands Four pair of glands located on the lower abdomen of the honeybee, one pair each under the third through sixth abdominal segments, which function when the abdomen is full.

Wax moth Several species of moth, whose larvae feed on the wax combs of the honeybee.

Wild hive Any hive of bees not housed in a standard hive in which the frames cannot be moved or manipulated.

Worker bee The incomplete female honeybee whose caste makes up the bulk of the population of a hive and who perform the many tasks necessary for the survival of the colony.

INDEX

Absconding, 59
    why bees do it, 126
AFB/EFB
    also see American Foulbrood
    characteristics of, 41
    control of, 39
    description of, 38
    dispensing (illus), 64
    fall medication to control, 239, 241
    how it spreads, 39
    medicating to control, 100
Africanized bees, 319
Afterswarms, 124, 125
Alarm, honey overflow, 222
Almanac yearbook, 4
American Foulbrood, 4
    also see AFB
    medicating to control, 60, 61
    signs of, 58, 59
Auxiliary hive exit for winter, 251

Bagging honey supers, 233
    (illus), with Paradichlorobenzene, 234
Bait hive, bees swarming into, 121
Ball of bees, 131
    if you find a, 86
Banking queens, 172
Bee blower
    how it works, 203
    (illus), 203
    use of, 204
Bee diseases
    booklet to obtain, 29
    description of the, 196
    diagnosis of, 42, 59
    how to spot, 38, 41

Bee escape
    how to use the, 196
    (illus), 198
    possible problems with the, 199, 200
    use of the, 197
    when to use the, 196
Bee-Go, 201
Beehive
    first year, 19
    producing, 19
    producing (illus), 19
    second year, 19
    spring cleaning of, 56, 57, 58
    spring condition of, 56, 57, 58
Beekeeper's
    clothing, 43-46
    clothing (illus), 44
    coveralls, 44, 45
    veil, 44, 45, 46
Beekeeping
    association, 4
    climates, 3
    supplies, ordering, 26
Bee sting, relief from, 54
Bee stinger
    how to remove, 53
    removal (illus), 54
Bees
    brushing, 194
    feeding a package of (illus), 97
    feeding a package of, 96
    how to gorge the package, 97
    how to inspect package, 95
    in houses, 268
    in odd movable hives, 274
    in storage area, 202
    in trees, 268

installing a package of, 95
mean, 287-289
methods of installing package, 99, 111
ordering packages of, 13
orienting, 290
package of (illus), 12
scenting, 148
scenting (illus), 292
shaking from frames, 132
shipment of packages, 12
short history of, 318
size of package to order, 100
to install more than one package, 99
when to order, 7
when to medicate, 7
why gorge the package, 97
Beeswax, melting temperature, 210
Benzaldahyde, 201
(illus), 202
Blower velocity needed, 203
Boot bands, 45, 46
Borrowing brood frames, 263
Bottling honey, 228
Bottom supering, 180, 188
Boxed swarm (illus), 149
Brace comb, 72, 73
Branch, swarm on a thick, 150
Brood chamber
   first inspection, 116
   how to inspect, 77
   inspection of, 71, 73
   reversing the, 126, 129
   what to look for in the, 71
   when to inspect the, 71, 117
Brood frame, what to look for on a, 73
Brood frames
   borrowing, 263
   moved into the honey super, 129
Brood inspection, reasons for, 71
Brood pattern, unusual, 262
Brood super, reversing, 139
Brood supers, when to inspect, 82
Brood
   bald or bare headed, 78
   chilled, 176
   how to check frames of, 78
   moving it above the queen excluder, 136
   uncapped, 78

Brushing bees, 194
   how to, 195
Burr comb, 73, 197
   between frame tops, 72
   cutting from between frame tops, 73
   cutting from between frame tops (illus), 74

Cage
   home made queen, 158
   see queen cage
Calendar of manipulations, 7
Candles, melting wax for, 316
Candy plug
   perforating the (illus), 89
   why not perforate, 89
Cappings basket, 224
   how to make a, 224
Cappings scratcher
   how to use the, 220
   (illus), 220
Cappings, handling of, 222
Chemical bee removal, 200
Chunk honey, 208
Cleaning extracting equipment, 228
Clipped wing queen, 126, 142
Cluster
   movement of in winter, 178
   the winter, 239
Cold honey, 215
Comb boxes
   foil, 207
   plastic, 207
Comb building, 181
Comb honey
   cutting, 207
   in frames, 208
   in section boxes, 206
   liquefying, 210
   storage of, 207
Comb
   brace, 72
   burr, 72
Coveralls, 44, 45
Cream of tartar, 65
Crosscombing, 183
Crowding, causes swarming, 126

De-queen, when to, 81
De-queening

a swarm, 160
   instructions for, 83
   why it is necessary, 125
Dead hive
   restocking a, 12
   storing a, 236
Demonstration hive, see observation hive
Divides
   how to make, 105, 132
   when to make, 131
   where to locate, 133
Drone cells, capped (illus), 261
Drone
   dies on mating, 122
   his importance to the hive, 122
Dumping bees
   how to, 85, 86, 87
   to find the queen by, 85
   why it works, 85

EFB
   also see European Foulbrood
   characteristics of, 41
   control of, 40
   description of, 40
Eggs
   how many the queen lays, 123
   (illus), 79
   multiple in cell, 77
   pattern in the frame, 117
   several in a cell, 259
   two or more in one cell, 73
Emergency feeding, 254
Entrance, when to reduce, 256
Equipment, where to buy, 21
European Foulbrood
   also see EFB
   medicating to control, 60, 61
Evaporative cooling of hive, 67
Extracted supers, storage of, 232
Extracting equipment, cleanup by the bees, 229
Extracting, 213
   honey, 218
   in mid-flow, 191
   out of doors, 217
   preparing for, 214
   setup (illus), 225
   when to begin, 213
Extractor draining, cautions against overflow, 222
Extractor
   radial, 215
   spinning time, 221
   tangential, 215
   turning speeds of, 221

Fall feeding of bees, 243
Farmers' Bulletin #2255, 4, 29, 55, 78
Feeding
   emergency, 254
   in the fall, 243
   internally, 242, 243
Foulbrood Disease, symptoms of, 58
Foundation, chewed away by bees in super, 180
Frame
   cleaning tool, 209
   grip, use of a (illus), 74
   how to inspect a, 73, 78
   how to turn while inspecting, 76
   manipulation, 180
   removing for inspection, 179
Frames, shaking to find the queen, 85
Fuels for smoker, 48
Fumadil-B
   how to use, 65
   see Nosema Disease
Fume board, 200
   how it works, 201
   (illus), 202
Fume pad, 200

Gate, honey, 225
Gloves, 44, 45, 46
Gorging bees, how to, 97
Gorging, by bees preparing to swarm, 120
Granulated comb honey, 209
Granulation of honey, how to prevent, 205

Hive
   after the swarm leaves the, 121, 122
   bees cool using water, 67
   bomb (illus), 47
   cooling, evaporative, 67

INDEX 331

during the honey flow, 177
entrance screen, (illus), 52
first year (illus), 25
how to enter, 49
how to move, 51
in midwinter, 28
locating the, 50
moving a, 52
population sizes, 178
queenless, 125
ramp to front of, 51
ready to move (illus), 52
reasons it may die, 35
staples (illus), 52
starvation, causes of, 178
temperature inside, 35, 67
tool, one use of (illus), 74
venting, 178
when to move, 51
when to check the, 117
winter dangers, 35
wrapping for winter, 252, 253
Hives
    changing locations of, 130
    dampness in winter, 36
    dividing, 128
    increasing the number of, 128, 132, 133
    moving for swarm control, 137
    natural enemies of, 124
    reversing entrance of, 133
    starting new, 24
    switching locations of, 137
    when to medicate, 7
    when to unite, 8
    wild, 72
Hiving a swarm, 152
Honey bottling, 238
Honey flow
    description of, 175
    duration of, 175
    in the North, 191
    in the South, 191
    winter, 214
Honey gate, 225
Honey overflow alarm, 222
Honey spinner, see extractor
Honey storage, supers for, 23-27
Honey strainer (illus), 223
    pantyhose used as a, 223
Honey straining, 216

Honey super
    adding the, 187, 189, 192
    inspecting the, 186
    rule of thumb for adding a, 185
    storage, 233
    using nine frames in, 183
    when to remove a, 181
Honey supers
    adding, 179
    capacity of, 19
    selecting, 20
    ways to remove, 194
    weight of when filled, 19
    when to add, 158
    when to remove, 194
Honey
    bees watering down, 33
    chunk comb, 208
    cold, 215
    draining from extractor, 227
    granulation of, 209
    how to strain, 217
    thick or sticky, 215
    types produced (illus), 21
    types produced, 20
    when to extract, 191, 193
    yield per super, 217
Honeycomb
    drawn by the swarm, 121
    how it is made, 120, 121
Hornet sting, 54

Inner cover
    screened (illus), 198
    use of the (illus), 198
Inspecting package bees, 112, 113, 115
Inspection notes, keeping and using, 192
Inspection of brood, 73
    (illus), 75
Insurance on package bees, 95

Knife
    electric uncapping, 219
    uncapping (illus), 219

Ladder, extension, 151
Larvae, (illus), 79
Laying pattern, 73

332 INDEX

Laying worker brood, 261
  disposal of, 263
Laying worker hive, 125
  re-queening the, 260
Laying workers, 125, 259
  signs of in the hive, 77, 259
Liquefying honey, 210, 212
  temperatures for, 210, 211

Man-made swarm, 129
  how to make, 131
  when to make, 131
Mating flight
  description of, 122
  of the queen, 122
Mean bees, 128, 287-289
  how to de-queen, 288
Medicating hives, 63
Medication, fall, 239
Melting honey, 210
Mice, in winter, 254
Mite, Varroa, 319
Moisture in the hive, 253
Moth, see wax moth
Moving hives, steps for, 53
Multiple eggs in a cell, 259
Multiple swarms, 124

Nectar flow, definition of, 175
Nine frames
  in the honey super, 183
  starting with only, 183
Nosem-X, used to treat Nosema Disease
Nosema Disease
  control of, 42, 242
  description of, 41
  how to treat, 65
  medicating against, 100, 112, 116
  results of, 41
  storing medication for, 66
Nosema-medication, how to mix the, 65
Notes, inspection, use of, 192

Observation hive, 297
  activities you can see, 304-308
  size recommended, 298
  starting with a frame of brood, 299
  starting with a small swarm, 301
  starting with a swarm (illus), 301
  swarm prevention in an, 303
  travel tube of the, 302
Odd movable hives, 273
Orientation flights, 290-291
Ovaries of queen, 122
Oxytetracycline hydrochloride, 61

Package bees
  critical time for, 100
  feeding, 96
  four day inspection of, 112, 113, 115
  gorging, 97
  how to inspect, 95
  installing (illus), 101, 105, 109
  installing wet, 107
  Method 1 for installing, 100
  Method 2 for installing, 104
  Method 3 for installing, 107
Parthenogenesis, 122
PDB, 182, 232
Pesticides, danger to bees, 124
Pipping, by the queen, 124
Pollen
  for human consumption, 313
  gathering for bee food, 313
  when to trap, 62
Pollen patty
  giving bees a, 112, 116
  (illus), 64
  preparation of, 63
  when to use the, 62
Pollen substitute, 62
Pollen supplement
  how to store the, 63
  how to use the, 63
  ingredients in, 62
  mixed with terramycin, 63
  patty, 112-116
  why feed, 35
Pollen trap
  as part of hive (illus), 312
  hanging from the hive (illus), 311
  how it works, 309
  types of, 310

uses for, 62
Pollen traps, 309
Pollination, 175
    by the wind, 176
Post, swarm on a, 149
Premedication check, 55
    how to make the, 55, 56
    when to make the, 55
Preparing the hive for winter, 239
Pupae, how to uncap a few, 78

Queen cage
    cardboard, 93
    cardboard (illus), 92
    descriptions of, 99
    hanging on frame (illus), 103
    homemade, 159
    in package of bees, 99
    on brood comb (illus), 92
    on frame tops (illus), 90
    plastic, 99, 102, 110
    putting in hive, 90, 91
    removing from package, 98, 108
    reusing a, 159
    types of, 98
    wooden (illus), 103
    wooden, 88, 93, 99, 102, 110
Queen cages (illus), 12, 16
    types of, 16
Queen catcher, 160
    (illus), 159
Queen cell cups, 141
Queen cells
    finding, 140
    how to tear out, 141
    (illus), 141
    tearing out, 130, 137, 140, 141
    when to tear out, 130
    where to find, 130, 141
Queen excluder, 185
    moving brood above the, 136
Queen rearing, 282-286
Queen scent, 125
    attracts drones, 122
Queen substance, 125
    lack of, 125
Queen succession, 123
Queen
    aging of, 126
    capturing from a swarm, 158
    dead in package, 111

    do you have a, 76
    dumping bees to find the, 85
    fertilization of, 122
    flying from the hive, 119
    how to cage the, 159
    how to catch, 159
    how to find on frames, 85
    how to get her alone, 82
    how to pick her up, 82
    if dead in package, 111
    introduction of, 88
    laying pattern of, 117
    missing from hive, 125
    none in the hive, 259
    releasing attendants from cage, 82
    replacing dead, 111
    swarm, 169
    virgin, 259
    when and how she mates, 122
    when to check for the, 81
    with a clipped wing, 126, 131, 142
Queenless hive, 125, 244
Queens
    banking, 17, 172
    for fall re-queening, 17
    for midseason re-queening, 17
    ordering, 15
    perils for the virgin, 123
    two in the hive, 123
    where to order, 17

Radial extractor, 215
Ramp
    placement of, 51
    to hive entrance (illus), 156
    why use a, 51
Re-queen
    how to, 89
    when to, 81, 132
Re-queening
    in the fall, 239-240
    prevents swarming, 126
    some reasons for, 126
    when and why, 126
    why it is necessary, 81
Removing bees
    from houses, from trees, 268
    from odd movable hives, 281, 282

(illus) from an odd hive, 278
Reversing brood chambers, 139
Robbing
    description of, 293
    emergency measures to stop, 295
    frantic, 294
    how to stop, 294
    leisurely, 293
    prevention of, 294
Royal jelly, 119
    food for the queen, 122
    tasting, 141
Rubber band, used on the frame grip, 75
Rule of thumb, 179
    for adding supers, 117, 185

Saving wild comb (illus), 278
Scrap wax, 197
Scratcher, cappings (illus), 220
Screened inner cover, 196
Settling tank, 224
    (illus), 226
Shaking bees, why it works, 85
Skunks, 72
Smoker alternative, 47
Smoker fuels, 48
    how to light bulk fuels, 49
Smoker
    how to light, 48
    (illus), 47
    sizes and types of, 48
Smoking the hive, 46
Solar melter (illus), 315
Sperm, queen's supply of, 122
Spermatheca of queen, 122
Spring activity
    first signs of, 33
    preview of, 33
Stacking honey supers
    (illus), with Paradichlorobenzene, 235
    with Paradichlorobenzene, 233
Staples, hive, 52
Starvation of hive, 57
Starving bees, how to feed, 57
Starving hive, how to save, 57
Stinger, how to remove, 53
Storage of honey in freezer, 206
Strainer (illus), 223

Straining honey, 216
Strong hives, moving, 137
Succession of the queen, 123
Sugar solution
    how to mix, 65
    preparing, 96
    recipes for, 65
Super, honey yield in the, 217
Supering
    bottom, 180
    top, 180
Supers
    adding honey, 184
    bagging for storage, 233
    cleaning and storing, 230
    how to clean-up sticky, 231
    reversing brood, 139
    Rule of thumb for adding, 117
    when to remove, 194
Support pins, (illus), 198
Swarm control
    description of methods for, 126-131
    manipulations, 127-131
    planning for, 8
    when to start, 131
Swarm queen, 169
    removing from the swarm, 164
Swarm queen cells (illus), 141
Swarm uniting
    methods, 163
    Carrier's method, 163, 165
Swarm
    before you capture, 146
    capturing (illus), 145
    capturing in a box (illus), 145
    description of, 118, 291
    dumping (illus), 165
    dumping to capture the queen, 158
    equipment used to capture a, 147
    front dumping of, 164
    hiving a, 152
    how to capture a high, 151
    if you lose a, 143
    in a capture box (illus), 149
    instructions for capturing a, 147-151
    man-made, 129
    on a high supple branch, 151

on a large branch, 150
on a post, 149
on a solid object, 149
on a tree trunk, 149
on the ground, 148
removing the queen from the, 164
uniting with paper, 167
vulnerability of the, 124
why uniting works, 167
Swarming
an irreversible process, 118
causes of, 126
discussion of, 118
how to prevent, 126
length of season, 118
reasons for, 118
signs of impending, 120
Swarms
capturing and hiving, 144
de-queening, 160
how to use, 161, 169-171
(illus), 119
inspection of man-made, 134
medicating, 153, 155, 157
multiple, 124
times they occur, 121
uniting man-made, 135
where to find, 144
which bees go with the, 120
Sweat band, 45, 46

Tangential extractor, 215
Tank, settling (illus), 226
Temperatures, 4
Terramycin (illus), 61
medicating with, 112-116
Terramycin mix, 157
how to dispense the, 63
how to prepare, 61
how to store the, 62
Top supering, 180, 188
Transferring wild comb to save the brood, 274

Uncapping honey combs (illus), 219
Uncapping knife
electric, 219
(illus), 219
Uncapping tools, 216
United hive (illus), 245

Uniting hives, 244
Unusual brood pattern, 262

Varroa mite, 319
Venting the hive, 178, 251

Wax
melting for candles, 316
scrap, 197
to produce extra, 183
trading for foundation, 317
Wax flakes, secretion of, 181
Wax melter, see solar melter
Wax moth
behavior of, 265, 267
causes of, 265
control of, 265, 267
damage from, 244, 264, 267
signs of, 266
to destroy, 265
When to extract, 213
Why hives starve, 251
Why unite hives, 244
Wind pollination, 176
Winter
feeding, emergency, 254
hive needs, 251
honey flow, 214
venting, (illus), 255
Wintering hives
in the South, 255
covering with a special box, 252
Workers, laying, 259